Enterprise Mobility

TECHNOLOGY, WORK AND GLOBALIZATION

The Technology, Work and Globalization series was developed to provide policy makers, workers, managers, academics and students with a deeper understanding of the complex interlinks and influences between technological developments, including information and communication technologies, work organizations and patterns of globalization. The mission of the series is to disseminate rich knowledge based on deep research about relevant issues surrounding the globalization of work that is spawned by technology.

Enterprise Mobility

Tiny Technology with Global Impact on Work

Carsten Sørensen

First published 2011 by
PALGRAVE MACMILLAN

Palgrave Macmillan in the UK is an imprint of Macmillan Publishers Limited, registered in England, company number 785998, of Houndmills, Basingstoke, Hampshire RG21 6XS.

Palgrave Macmillan in the US is a division of St Martin's Press LLC, 175 Fifth Avenue, New York, NY 10010.

Palgrave Macmillan is the global academic imprint of the above companies and has companies and representatives throughout the world.

Palgrave® and Macmillan® are registered trademarks in the United States, the United Kingdom, Europe and other countries.

ISBN 978–0–230–23607–3

This book is printed on paper suitable for recycling and made from fully managed and sustained forest sources. Logging, pulping and manufacturing processes are expected to conform to the environmental regulations of the country of origin.

A catalogue record for this book is available from the British Library.

Library of Congress Cataloging-in-Publication Data
Sørensen, Carsten.
Enterprise mobility : tiny technology with global impact on work / Carsten Sorensen.
p. cm.
Includes index.
Summary: "There are currently 3.5 billion mobile phones in the world and mobile information technologies permeate all aspects of life. This book explores how mobile technologies and information work shape each other. Most writings do not consider how information work increasingly relies on mobile services; this book seeks to address this imbalance" — Provided by publisher.
ISBN 978–0–230–23607–3 (hardback)
1. Mobile communication systems—Social aspects.   2. Information technology—Social aspects.   3. Communication and culture.   I. Title.
HM1206.S6567 2011
303.48'33—dc23   2011028843

10   9   8   7   6   5   4   3   2   1
20   19   18   17   16   15   14   13   12   11

Printed and bound in Great Britain by
CPI Antony Rowe, Chippenham and Eastbourne

To Nora

Contents

List of Figures and Tables

Figures

Preface

My parents worked in retail from before I was born. My father managed a small co-operative grocery store in the windswept north of Denmark, where I worked after school from the age of eight. When I was around 10, I encountered an alien and fascinating technology. In the late 1960s, punch-card technology was introduced to order supplies from the central warehouse (Heide, 2005). As a result, our little store received piles of empty punch cards with perforated chaffs, a special device to hold a card and guide the process of punching out chaffs, and a metal pen to do the job. Computerisation was at the edges of this strictly paper-based process. A shop assistant armed with cards, the holder and pen would order items by walking around the store and punch in data, and the cards would subsequently be picked up and fed into a central computer. By adapting the human operator to the requirements of the computerised replenishment system, the time and costs involved were greatly reduced compared to before, where orders were made either on paper forms or over the phone, requiring duplicate data entry and leading to more errors. The alien technology quickly became part of everyday life, the store got its own small punch-card reader connected to a telephone modem, and the replenishment process continued to be further standardised and streamlined.

Lesson 1: information technology can greatly standardise and streamline routine activities, even at the remote edges of organisations.

When I was in seventh grade, as part of preparing for working life, I spent a relatively boring and uneventful week in the data centre for a cement factory. However, as beautifully described by Richards and Alderman (2007), the blinking lights on the IBM mainframe were all etched into my brain. Five years later, studying mathematics and computer science at the local university just down the road from the data centre, I learnt another lesson on dramatic technological change. The established career path for computer professionals had up to that point involved a short 18-month education as programming assistants. However, the advent of the personal computer and the increased complexity of applications meant that the education was closed down because of a 70–80 per cent unemployment rate of new graduates. Employers demanded more abstract skills and these required a more comprehensive university education.

Lesson 2: follow the rapid technological development.

I wrote my MSc dissertation on computer science in 1988 on a little-researched phenomenon of the organisational introduction of standard application packages. Standard application packages were generally considered to be far inferior to bespoke software – the programmers union used the derogative term 'plastic software'. However, it was obvious to me then that significant gains could be made from buying a software product instead of hiring a group of systems developers. Whereas software production during the mainframe computer era was defined in terms of a process, the personal computer era focused on shrink-wrapped applications and market exchange of products. Now, software is increasingly considered as services available as and when needed. Generative platforms with low barriers to entry merge software production with global crowdsourcing, resulting in some software offered at radically lower price than before.

Lesson 3: as the result of industrial production and bespoke craftsmanship, software is generally produced for the many and then appropriated into specific contexts.

After having given a talk in London several years ago, a middle-aged gentleman approached me, sharing his experiences as one of the first mobile phone owners in the UK in the early 1980s. The bulky phone and running costs had set him back tens of thousands of pounds, but he added that 'it was the single best investment I have ever made in my life'. He was the CEO of a small demolition company and therefore makes his living from knocking down old buildings. Those with such needs generally want the best firm for the job and need to trust that all will go well. This means that providing direct access for prospective clients is the single most important aspect of driving the business. Through the mobile phone he nurtured direct client contact and as a result he landed significantly more jobs. Today everyone has a mobile phone and the challenge is no longer how to reach each other but rather to intelligently cultivate how, when and why to interact with others. Unfortunately, while technological development has raced on and produced a plethora of new gadgets and services, we still have not properly understood how to deal with this particular problem.

Lesson 4: mobile connectivity profoundly changes human interaction and decision-making.

The primary purpose of this book is to explore the diversity of mobile services by studying how it is used across organisations. Therefore, the book neither provides a purely technological account nor does it exclusively focus on understanding organisational changes under mobile working.

There is an emerging field of research studying the social impact of the mobile phone. The technology studied in this book is not only a physically tiny technology, it has been offered relatively little research attention. Having looked high and low, I have, however, yet to discover a research monograph on enterprise mobility, but I hope this will be the first of many such books.

I hope you will enjoy reading the book as much as I have enjoyed writing it!

Carsten Sørensen

London, February 2011

www.carstensorensen.com

mobility.lse.ac.uk

beingmobile.org

Acknowledgements

Reversing the tradition of always mentioning the most important at the end, a huge thanks to my lovely wife Nora for sacrifices through the writing of this book, and to little Anna who in the process learnt how to challenge boundaries between work and play through stealth and humour. I am grateful for my parents always instilling in me enough self-confidence to play with the big boys and girls. A big thank to the family in Baitâ who always look very well after me. In May 2010, I sadly lost my brother Morten, who in 1996 took part in a study of mobile email and GPS in his lorry company.

A special thanks to all the mobility@lse (mobility.lse.ac.uk) doctoral students who directly contributed to this book through their fieldwork (in chronological order): Masao Kakihara, Daniele Pica, Adel Al-Taitoon, Gamel Wiredu, Jan Kietzmann, Silvia Elaluf-Calderwood and Kofi Boateng. Also a big thank you to all the anonymous participants who let us into their mobile working lives. Special thanks to Daniele Pica for helping debug the arguments and being a good friend throughout. Also a big thank you to Lars Mathiassen for essential advice at a critical stage of the project and for always being there.

It all began in Gothenburg, 1993, with Fredrik Ljungberg and Henrik Fagrell's doctoral work as part of the Internet Project and continued in Uddevalla at the Laboratorium for Interaction Technology 1996–2006. Many people have offered their help and I hope I have not forgotten anyone. Thanks to Christian Licoppe, Kalle Lyytinen, Christian Heath, Urban Nuldén and Steve Sawyer, Ralph Schroeder, Leopoldina Fortunati, Allen Lee, Alfred Waller, Jonathan Ezer and Pål Sørgaard. Thanks to the book series editors, Leslie Willcocks and Mary Lacity, and to Stephen Rutt and Eleanor Davey Corrigan from Palgrave Macmillan. Thanks to Jon Lloyd for excellent copy-editing. My colleagues at ISIG, LSE, have kindly supported my research over the years. Also thanks to all the doctoral and MSc students I have engaged with over the years.

Microsoft has kindly provided support for some of the work through the Tomorrow's Work Programme, as has Google through their Future of Work project. Rob Gear and David Elton from PA Consulting have contributed with interesting discussions on innovation.

The end result is of course entirely my own responsibility.

1
Mobility – Emerging Challenges

There are over five billion GSM mobile phone connections globally and a growing number of other mobile information technologies permeate all aspects of life.[1] The number of mobile phone subscriptions in developing countries (non-OECD members) was in 2008 estimated to be three billion out of the then four billion total subscribers, dwarfing the global total of two billion Internet connections in 2010 (Kluth, 2008; *The Economist*, 2009). The mass diffusion of miniaturised computers linked together in personal, local and global networks has created unprecedented technological intimacy over global infrastructures. Mobile information technology touches an increasing proportion of our human existence. Whether at home in bed, on holiday on a beach or at work at the desk, we have instant connectivity. Of all such technologies, the mobile phone makes up by far the largest and most visible single category. So far, an emerging research field has explored the general social impact of this single technology, the mobile phone, while very little research has investigated *enterprise mobility* – the application of diverse mobile information technologies in the context of work (Barnes, 2003; Basole, 2008).

This book is the first comprehensive monograph to explore enterprise mobility. The aim is to understand the diversity of mobile services portfolios and their role in mobile work. The book presents enterprise mobility as balancing the creation of fluidity with the management of boundaries in relation to the three classes of mobile activities; interaction, collaboration and control. This chapter discusses; the context for the book, its contribution, related research, research approach and readership.

1.1 A world of mobility

It is estimated that in 2013 there will be more than 1.1 billion mobile workers globally, representing 35 per cent of the total workforce (Ryan *et al.*, 2009). Providing mobile access to enterprise systems and to further develop enterprise mobility in the organisation is a key priority for 3,000 IT decision-makers in a 2010 survey, which also indicated that 75 per cent of organisations deploy mobile information technology to improve worker productivity (Sheedy, 2010). The vast majority of workers already now possess the opportunities for mobile interaction through mobile phones and Wi-Fi-connected notebook computers. One survey shows 86 per cent of companies providing mobile email access (Sheedy, 2010). A growing group of 'digital nomads' spend one-third of their time in traditional offices, one-third of their time in home offices and the remaining third working from public places, cafes or similar transitional places (Kluth, 2008; Rosenwald, 2009).

Twenty-first-century information work

Societies are not what they used to be. Long ago, manufacturing replaced the primary sectors of agriculture, forestry and fisheries as the primary force. During the 1960s and 1970s the notion of the post-industrial society, or service society, provided a view into the current situation in most industrialised countries (Halmos, 1970; Gersuny and Rosengren, 1973; Bell, 1976). While the service society relies more on manufactured goods than ever before, the tertiary sector still employs three-quarters of the workforce in industrialised countries (*The Economist*, 2007). Globalisation, market demands and technological development place the production of knowledge and innovation at the centre of concern and are equally fuelled by them. Contemporary life signifies a 'liquid modernity' where erstwhile stable boundaries and clear frames of reference are disappearing, leaving individuals to pragmatically plan under conditions of uncertainty and constant change (Bauman, 2000). Industrial society's well-defined understanding of space and time is replaced with a space of flows fuelled by societal activities extensively organised through computerised networks (Castells, 1996).

Organisations are not what they used to be. The macro-economic developments of the past century place organisations in situations characterised by paradoxes, tensions and competing requirements concurrently requiring, for example: operational improvements and strategic innovations; internal and external focus; and the need for both increased organisational control and flexibility. Individuals are expected to support the organisation in resolving such tensions by enabling it to become ambidextrous – enabling concurrent short-term optimisation and long-term innovation (O'Reilly and Tushman, 2004). However, relying on organisational structures to provide such ambidexterity is less effective than devolving it to individuals and groups, thereby enabling contextual ambidexterity through combinations of planned interventions and emerging decisions (Birkinshaw and Gibson, 2004).

Work is not what it used to be. Organisational pressures and opportunities have reshaped working life. An increasing proportion of working arrangements deviate radically from the traditional 9-to-5 job shaping a 40-year-long career with one employer. Complex arrangements of itinerant workers, contractors and free-lancers spanning projects, teams and organisations reshape organisational life (Castells, 1996, pp. 281–9; Barley and Kunda, 2004). Flexible working arrangements, the intensification of work and pressures to meet a range of conflicting demands are the order of the day for a growing proportion of the working population (Carnoy, 2002; Bunting, 2004; Felstead *et al.*, 2005). Demands for the advanced management of information and co-ordination of efforts require the application of a variety of information and communication technologies in most workplaces. Whereas in the past interpersonal interaction at work was generally considered to be counter-productive, it is now central (Conradson, 1988). In the context of this book, information work characterises the aspects of work pertaining to interaction and information management. Such information work forms an integral part of almost any job and very few workers do not rely on information technology.

A survey of 3,000 information workers across France, Germany, Japan, the UK and the USA demonstrates a modern workforce with an articulated understanding of the critical importance of flexibility, collaboration and innovation (Future Foundation, 2010). The respondents voice a strong conviction of a direct link between an organisation's ability to collaborate and innovate. They place significant importance on the role of information and communication technology as the enabler of such efforts. Seventy-one per cent of the respondents regularly contribute with ideas, 60 per cent have seen these successfully implemented and 31 per cent believe that future innovation will be driven by freelancers and external ideas consultants (Future Foundation, 2010, p. 9).

Everywhere to go and nowhere to hide

Flexible working practices have taken many forms, for example: telecommuting, home-working, shared offices, hot-desking, global virtual teamworking and mobile working. Such new working arrangements respond to changing demands and signify new ways of conducting, co-ordinating, managing and measuring work. The changing practices serve as a critical element in the organisation, achieving contextual ambidexterity through resolving organisational paradoxes by rendering workers both more independent and increasingly mutually inter-dependent (Castells, 1996; Dubé and Robey, 2009).

Mobile work is the most radical form of flexible working. It has the characteristics of local geographical mobility, where the worker moves within a restricted local area, and remote working, where workers collaborate while away from fixed workstations. Enterprise mobility engages the mutual development of mobile work and mobile information technology practices in very large and very small organisations. The technology alters the possibilities for the organisation and management of work. These changes in turn provide new opportunities for the application of innovations in mobile information technology. Enterprise mobility raises issues far beyond the simple application of the mobile phone at work for restricted groups within organisations. It is in principle a matter for any worker and not only the mobile executive elite or the traditionally mobile roles, such as salespeople, repair engineers and delivery drivers.

Technology is not what it used to be. The mainframe helped organisations to streamline administrative processes. The personal computer improved productivity, for example, by helping individuals make local sense of complex enterprise data and subsequently make decentralised decisions based on this understanding. Enterprise mobility relies on the dramatic technological revolution miniaturising and networking the computer in personal, local and global networks. Mobile computing challenges the assumptions of how work is arranged and how it is understood as interaction between individuals. Enterprise mobility uniquely relies on the cultivation of intimate user-technology relationships in everyday routines and practices, and on new emerging forms of interaction.

It is tempting to assume that distance does not matter and that work can be conducted at any time and anywhere through combinations of global networking and mobile information technology. However, this conclusion makes inappropriate assumptions of both the power of technology and the inherent characteristics of work. Paradoxically, technology both provides the potential for organisations to achieve detailed control over mobile activities as well as extensive opportunities for individuals to engage in emerging decisions in order to meet unfolding demands. This paradox is a source of mobile contextual ambidexterity. Technology performances intelligently appropriate technological opportunities for action within the context of organisational contradictions, conflicting demands and paradoxes. Emerging needs for decisions can be met with informed and co-ordinated action. Key aspects of the corporate infrastructure can be placed directly in the hands of mobile workers both at the centre and at the periphery of the organisation. Enterprise mobility affords not only great flexibility within the appropriate context but also significant opportunities for extensive surveillance. Modern workers have everywhere to go and nowhere to hide.

Cultivating fluidity and boundaries

For mobile workers a combination of planned and emerging technology performances shape their work; stipulated schedules may be kept while exceptions encountered can result in interaction improvised on the spot. Mobile information technology provides widespread opportunities for challenging boundaries through fluid interaction. An interactional cacophony emerges from instant connectivity through mobile phone conversations, SMS messages, mobile email, mobile instant messaging, Facebook updates, Twitter postings, etc. Mobile information technology has introduced the possibility for unrestricted synchronous and asynchronous connectivity wholesale. The use of the technology therefore introduces the challenge of managing this fluidity as an essential resource for mobile workers. Fluid interaction represents the advantage of instant decisions while placing demands on individual attention, breaking through borders shielding collaborators from being disturbed and challenging carefully crafted organisational boundaries and processes.

The fluidity of modern life and work does imply the delegation of the creation and maintenance of boundaries to individuals and groups. When work was exclusively characterised through the boundaries of co-present interaction around fixed workstations, re-organisation exclusively marked a process of physically re-arranging people and workstations. Such re-organisation can through enterprise mobility be the result of creative technology performance by individuals or groups. Boundaries can, for example, form barriers against unwanted interaction requests, redirect interaction requests from synchronous to asynchronous media, create protected spaces for collaboration, streamline mobile information decision processes and create flexibly reconfigurable organisational boundaries.

The use of mobile information technology at work represents the everyday meeting of the paradoxes of mobile working and the technological opportunities for decisions through planned and emerging technology performances. Technological opportunities are here characterised in terms of mobile service portfolios representing a diversity of opportunities for managing information and interaction. An indepth exploration of mobile technology performance through nine case studies informs the discussion of the challenge at the core of enterprise mobility – the continuous cultivation of fluidity and boundaries. Figure 1.1 provides an overview of the proposition in the book – the cultivation

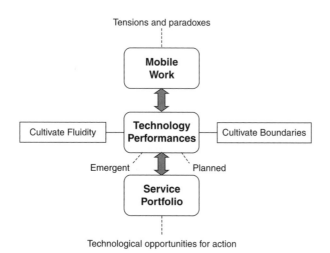

Figure 1.1 Overview of perspective on enterprise mobility

of fluidity and boundaries through emergent and planned technology perform-ance signified by the meeting of mobile work and technological opportunities.

This book provides unique insights into the diversity of technology perform-ances in mobile work. It situates the discourse on enterprise mobility within the context of organisational paradoxes and contextual ambidexterity, and pro-poses the key challenge in terms of the cultivation of fluidity and boundaries. Finally, it contributes to existing research by characterising the unique diversity of mobile services. This provides a taxonomy of six service categories defining enterprise mobility services in terms of: 1) *intimacy* in the user-technology rela-tionship; 2) *connectivity* services linking to remote others or information; 3) the *priority* of interaction; 4) *pervasiveness* through context-awareness; 5) the *memory* of ongoing relationships; and 6) the *portability* of service access.

1.2   Mobility research

A significant body of related research has potential relevance for this book, indeed far too much to comprehensively review here. Diverse academic fields such as social geography (Carlstein, 1983; Cresswell, 2006), sociology (Urry, 2007), philosophy (Lefebvre, 1991) and organisation studies (Dale and Burrell, 2008) explore the roles of space, place and mobilities in human activity. An emerging research field studies the role of the mobile phone and mobile com-munication in society (Ling, 2008). Studies of ubiquitous and urban computing explore pervasive, embedded and mobile technologies in order to understand the issues of user-computer intimacy (Dourish, 2001), computing and architec-ture (McCollough, 2004) and ubiquitous technology in social spaces (Bassoli *et al.*, 2007). Human-Computer Interaction (HCI) research studies the design of usable devices and services (Love, 2005), and a range of computer scientists and engineers explore the technical aspects of mobile technology innovation.

The following discussion focuses on social studies of mobile information technology, i.e., research with a keen interest in the mutual constitution of human and technical agency.

Social studies of mobile information technology

The social study of information and communication technology is a broad field drawing together a range of concerns and research communities interested in understanding the intricate relationships between technology, social actions and structures (see, for example, Avgerou *et al.*, 2004; Mansell *et al.*, 2007). Within this area a number of emerging research discourses explore mobile and ubiquitous information technologies. Classifying these into studies emphasising social life in general versus working life and into studies concerned with a single technology – the mobile phone – as opposed to a broader portfolio of mobile and ubiquitous information technologies, four clusters of research emerge: a) social studies of the mobile phone, exploring its use for social interaction; b) social studies of ubiquitous computing, considering a wide range of technologies; c) studies relating to the mobile phone at work; and d) enterprise mobility – the study of mobile and ubiquitous computing at work. Of the four, by far the most research has been conducted within strand a) – social studies of the mobile phone – with much less research being conducted on the other three. In particular, the organisational significance of mobile and ubiquitous information technology has not been met with equal research interest in enterprise mobility. The following sections briefly outline the four research strands.

Strand a): social studies of the mobile phone

The most significant body of socio-technical research in mobile information technology is concerned with one technology, the mobile phone, and with its use in general social contexts. A significant number of books, edited collections and journal articles comprise this emerging field studying mobile communication. With billions of mobile phones globally, it is not surprising that there is a perceived need to understand the mutual shaping of this new technology and emerging social practices. Indeed, it could be argued that this field is still in its early infancy and that there is still scope for a distinct theoretical agenda and discourse to emerge from the research. This research is at its core social science and displays little interest in understanding the diversity of technical opportunities.

For example, this research strand studies the history of the mobile phone (Agar, 2003), the co-evolution of technological practices and language (Baron, 2008), new communicative practices (Ling, 2004; Harper *et al.*, 2005) and changes in social rituals (Fortunati, 2002; Licoppe, 2004; Ling, 2008). This research also conducts regional analyses of the global impact of mobile telecommunications across countries (Castells *et al.*, 2007), in Japan (Ito *et al.*, 2005), in the Asia-Pacific region (Rao and Mendoza, 2004; Pertierra, 2007; Hjorth, 2009) and in developing countries (Horst and Miller, 2006; Donner, 2008).

Strand b): social studies of ubiquitous computing

While the mobile phone offers a strikingly successful form of information technology, everyday life is for most people partially shaped by what Yoo (2010) characterises as 'experiential computing', i.e., activities involving everyday artefacts with embedded computing capabilities – mobile phones, iPods, cameras, GPS navigators, digital photo-frames, toasters, ovens, etc. Compared with the significant interest in studying the mobile phone, little research has so far broadened the focus to include a range of mobile and ubiquitous information technologies – experiential computing.

For example, the social study of ubiquitous computing explores the role of technologies disappearing into everyday life in general (Weiser, 1991) and the theoretical framing of user-technology intimacy in particular (Dourish, 2001).

Researchers study the changing distribution of power and influence as a result of ubiquitous technology, for example, a shift towards corporate control over consumers with Radio Frequency Identification (RFID) technology (Albrecht and McIntyre, 2006) and, conversely, the role of citizen activism through wearable computing (Mann and Niedzviecki, 2002). Studies of urban computing have explored technology and the built environment in general (Mitchell, 2003; McCollough, 2004) and, within this, the role of public authoring (Angus *et al.*, 2008) and music sharing (Bassoli *et al.*, 2007).

Strand c): studies relating to the mobile phone at work

Compared with the significant amount of research on the mobile phone in a general social context, there is very little research on the use of the mobile phone at work. Given the importance of the mobile phone at work, this is striking. However, researching a single technology in the context of work is challenging as organisations tend to provide support through a range of mobile information technologies, the mobile phone is less and less associated with the isolated connectivity of voice calls and SMS messaging, and some organisations, such as the police, have decades of experiences with advanced portfolios of mobile technology. Therefore, maintaining a strict analytical focus on the single technology of the mobile phone at work raises the question of theoretical focus, as any subject studied is likely to rely on a portfolio of services across several information technologies.

Social studies of the mobile phone at work explore issues such as the role of the mobile phone in maintaining ongoing social contact whilst at work (Wajcman *et al.*, 2008) and in the intensification of work (Bittman *et al.*, 2009). Research has also been conducted into the role of another single technology, the personal digital assistant (PDA), in maintaining work-life boundaries (Golden and Geisler, 2007). Mobile email is a rapidly diffusing technology at work and the impact has been explored in Mazmanian *et al.* (2006) and Straus *et al.* (2010). Kellaway (2005) provides a highly amusing fictional account of the use of mobile email.

Strand d): enterprise mobility

Very few organisations have not been infiltrated by some mobile and ubiquitous information technologies, such as mobile phones, smartphones, notebook computers, tablet computers, barcode and RFID technology, and a variety of bespoke devises and systems. This led Lyytinen and Yoo (2002b) to outline a research agenda for the organisational study of mobile and ubiquitous computing.

A number of edited volumes investigate a variety of research issues related to enterprise mobility (for example, Sørensen *et al.*, 2005; Basole, 2008; Hislop, 2008). Practitioner books discuss the pragmatics of enterprise mobility, providing advice on the implementation of mobile working (for example, Lattanzi *et al.*, 2006; Darden, 2009; Reid, 2010). Although the transformation of work and flexible working played a significant role in Castells' (1996, pp. 216–354) analysis of the network society, the influence of mobile information technology on work is not given a similarly prominent treatment in his book on mobile communication and is only briefly mentioned (Castells *et al.*, 2007, pp. 78–83).

The social study of information and communication technology in organisations is at the core of the Information Systems (IS) field, yet this field (or indeed associated fields) has not been responsive to Lyytinen and Yoo's (2002b) call for action. In the ten-year period from 2000 to 2010, a total of 2,001 research articles were published in the basket of eight IS journals.[2] Of those a total of 76 (3.8 per cent) studied mobile information technology (Landau, 2010, p. 27). There is an inverse correlation between the journal impact factor and the propensity to publish articles on mobile information technology. The two highest-ranked journals, *MISQ* and *ISR*, have published two and four papers, respectively, on mobile information technology, while the lowest-ranked journal, *EJIS*, has published 25 articles. The four lowest-ranked journals have published 60 of the 76 articles on mobile information technology (78.9 per cent) (Landau, 2010, p. 32).[3] Comparing the social and organisational role of mobile technology, this does not seem to be a measured response from the IS research community, despite the slight increase in mobile information technology publications over the ten-year period studied.

1.3   Research approach

A significant diversity of research fields study mobile information technology in general by applying a diverse range of approaches (Love, 2009). For the study of enterprise mobility, any existing method discussion within IS can be replayed and appropriated, and current research into the social and organisational use of mobile information technology also illustrates a diversity of approaches: for example, technical design studies constructing new technology (Barfield and Caudell, 2001); field and laboratory experiments (Kjeldskov and Graham, 2003); analysis of industry data (Castells *et al.*, 2007); factor-based approaches for analysing survey data (Hong and Tam, 2006); qualitative interviews (Ling, 2004); and participant observation (Ljungberg, 1997; Wiberg, 2001; Horst and Miller, 2006).

Whilst existing research approaches can be applied to the study of mobile information technology, there can also be reasons to adapt methods to suit the specific character of the domain. Within human-computer interaction research it has, for example, been suggested that the unique characteristics of mobile information technology require new approaches to studying usability (Kjeldskov and Stage, 2004). Some of the research reported on in this book has applied specifically adapted data-gathering techniques, such as mobile note-taking and remote video recording.

Mobility@lse

The mobility@lse research unit within the Information Systems and Innovation Group, Department of Management, at the London School of Economics and Political Science (mobility.lse.ac.uk) has since 2001 studied the organisational use of mobile and ubiquitous information technology. A total of 16 doctoral research projects have been directly associated with this unit, of which 12 had been completed by early 2011. The core concern of the mobility@lse research is to understand the mutual constitution of mobile working and technological choices in developing, adapting and using mobile and ubiquitous information technology. This subject has been studied in depth across seven completed doctoral projects, of which two projects have studied more than one specific domain of work. Throughout the book nine cases of enterprise mobility will be described, analysed and discussed. These nine cases have been selected from the seven research projects and form the empirical foundation for the analysis in this book. The cases are throughout the book referred to as follows:

Case 1 – Hiro: a Tokyo CEO using and adapting a bespoke mobile phone (Kakihara, 2003).

Case 2 – Ray: a London Black Cab driver using multiple mobile phones and a computer cab dispatch system (Elaluf-Calderwood, 2008).

Case 3 – Khalid: a Middle Eastern mobile trader engaging with financial markets off-premises through a portfolio of mobile services (Al-Taitoon, 2005).

Case 4 – John and Mary: two UK police officers using a range of mobile information technologies when working in and around a response vehicle (Pica, 2006).

Case 5 – Simon: a UK security guard doing rounds using an RFID reader-enabled mobile phone connected to a central server (Kietzmann, 2007).

Case 6 – Winters: a member of a team of UK-based industrial waste management lorry drivers using RFID reader-enabled mobile phones connected to a central server (Kietzmann, 2007).

Case 7 – Jun: an itinerant Japanese town planner using a mobile phone and a notebook computer (Kakihara, 2003).

Case 8 – Jason: a London-based food delivery driver relying on mobile phone communication (Boateng, 2010).

Case 9 – Yin: a UK health service professional engaging in remote learning using a PDA (Wiredu, 2005).

These ten individuals signify carefully selected individuals from a large pool in the respective case studies. The research projects involved hundreds of individuals in total, which included at least 200 mobile workers, for example, over 60 Japanese mobile professionals, more than 30 taxicab drivers, 16 traders (of which eight were authorised for mobile trading), over 40 police mobile police officers, etc.

The empirical data in these nine cases was collected using a variety of approaches alone or combined, such as: qualitative interviews (Kakihara, 2003; Al-Taitoon, 2005; Elaluf-Calderwood *et al.*, 2005; Boateng, 2010), action research (Wiredu, 2005), participant observation (Pica, 2006; Kietzmann, 2007; Boateng, 2010); and remote in-vehicle video recording of activities (Elaluf-Calderwood *et al.*, 2005). Table 1.1 provides an overview of these cases, their research approaches, research areas and key publications for each case.

Background studies

In addition to the core research used in this book, a number of doctoral projects have been conducted in association with the mobility@lse unit and have informed the work: for example, the study of itinerant software developers in Greece (Voutsina, 2008) and a project studying the use of mobile information technology in a Brazilian bank (Saccol and Reinhard, 2006).

Over the years a large number of minor case studies of enterprise mobility have been conducted as part of MSc dissertation work within mobility@lse: for example, mobile support for UK travelling salespeople (Arora, 2003); the everyday work of UK financial services professionals cultivating ubiquity (Sørensen and Gibson, 2004); mobile services innovation in Brazil (Fontana and Sørensen, 2005); the use of mobile payment technology in Mumbai buses (Sarnobat, 2006); mobile auditor work in China (Qui, 2006); mobile information technology in technology consulting across Denmark, the Philippines, Saudi Arabia, Singapore, Sweden, the UK and the USA (Antero, 2006); and the use of mobile information technology supporting expeditions to South America and the North Pole (Oliveira, 2006).

From the rich variety of case studies listed above, spanning a number of different countries, it is clear that the research presented in this book is broadly informed, even if the selected nine cases only represent three different countries.

This book draws loosely on a variety of ideas previously published in journal articles and book chapters: for example, Chapters 2 and 3 draw partly on Kakihara and Sørensen, 2001; Kakihara and Sørensen, 2004; Tilson *et al.*, 2010; and Sørensen, 2011, Chapters 4, 5 and 6 relate to earlier ideas in a range of works (for example, Sørensen, 2004; Sørensen and Pica, 2005; Elaluf-Calderwood and Sørensen, 2008; Sørensen and Al-Taitoon, 2008; and Sørensen *et al.*, 2008) and Chapter 7 draws on ideas from Sørensen, 2010.

Data analysis

The analysis in this book engages in a second-order data analysis of the nine case studies in Table 1.1. However, it is not a traditional second-order data analysis for two reasons. Firstly, the author actively participated in the process of discussing the cases as data was collected and, secondly, the analysis will not revisit the primary data and will conduct an entirely new analysis.

Table 1.1 Nine cases of enterprise mobility

#	Worker	Year	Place	Method	Extent	Topic	References
1	CEO	2001–2002	Japan	Interviews	Cases 1 and 7 part of a study with 63 interviews and *in situ* observations	Mobilisation of interaction for modern Tokyo professionals	(Kakihara, 2003) (Kakihara and Sørensen, 2004)
2	London Black Cab driver	2004–2005	UK	Interviews and video observation	39 interviews and 14 hours of video-taped observations	The choice of location as a core business strategy and the role of mobile technologies in pooling resources and informing individuals	(Elaluf-Calderwood, 2008) (Elaluf-Calderwood and Sørensen, 2008)
3	Off-premises foreign exchange trader	2003–2004	Middle East	Interviews and observation	102 interviews plus observation of traders	Discretion and control in mobile working for off-premises foreign exchange traders	(Al-Taitoon, 2005) (Sørensen and Al-Taitoon, 2008)
4	Police officers	2002–2004	UK	Observation, interviews and focus groups	250+ hours participant observation with 40+ officers and managers, 20+ interviews, 2 focus groups	The rhythms of interaction with mobile information technology by operational police officers	(Pica, 2006) (Sørensen and Pica, 2005)
5	Security guard	2004–2005	UK	Action research	350 hours of meetings, interviews and observation for cases 5 and 6	Real-life experimentation with RFID-enabled mobile phone technology supporting new ways of working	(Kietzmann, 2007) (Kietzmann, 2008)

6	Industrial waste management worker	2004–2005	UK	Action research	350 hours of meetings, interviews and observation for cases 5 and 6	Real-life experimentation with RFID-enabled mobile phone technology supporting new ways of working	(Kietzmann, 2007) (Kietzmann, 2008)
7	Town planner	2001–2002	Japan	Interviews	Cases 1 and 7 part of a study with 63 interviews and *in situ* observations	Mobilisation of interaction for modern Tokyo professionals	(Kakihara, 2003) (Kakihara and Sørensen, 2004)
8	Delivery driver	2006–2007	UK	Observation and interviews	50+ people participating in interviews and participant observation	Establishing IT-mediated control of work tasks with low degree of discretion through enterprise infrastructure and mobile information technology	(Boateng, 2010)
9	Health professional	2002–2004	UK	Action research	Intensive collaboration with 16 students + others participating in project	Supporting situated and remote learning for medical professionals (Perioperative Specialist Practitioners) with mobile information technology	(Wiredu, 2005) (Wiredu and Sørensen, 2006) (Wiredu, 2007)

The analysis of the cases is a second-order analysis of the second-order constructs discussed by the fieldworkers in their respective case studies (Van Maanen, 1979, p. 541). I took active part in the original analyses, so the re-analyses of cases is more substantially informed than a purely theoretical analysis of existing research. The combined 1,783 pages of dissertations analyse a wealth of primary data and establish a range of theoretical insights concerning the organisational use of mobile information technology. The work of distilling key findings has been carried out for the duration of each individual research project, with the ongoing synthesis of findings forming a significant activity since 2004. The effort of characterising the diversity of mobile services intensified during 2009–10.

Although I participated to some extent in the collection of primary data in cases 4, 8 and 9, this was in a much smaller capacity than the primary researchers. I did, however, play an active role in the discussions of the analysis of findings in all the cases, and my role can therefore be characterised as slightly more detached and distanced in the analysis of the data (Prasad and Prasad, 2000, p. 391). In the context of organisation theory, it has been argued that such detached second-order analysis can sharpen the analysis of first-hand insights (Gioia and Chittipeddi, 1991; Kilduff *et al.*, 1997; Prasad and Prasad, 2000, p. 391). Furthermore, the aim of the analysis conducted here, as opposed to the original analyses, is both to reveal patterns of technology use across different enterprise mobility domains and to explore the diversity of mobile services applied as input to a general discussion of mobile services portfolios.

A significant proportion of the book is devoted to the three empirically dominated Chapters 4, 5 and 6. As the study of enterprise mobility is still in its infancy, there is a need to establish appropriate categories for analysis, and in particular to better understand the diversity of technological support and the relationships between such diversity and work. As argued by van Maanen (1979, p. 539):

> ... the amount of time an investigator spends constructing a theory by actively seeking the facts is a variable and one that presumably should be related to the quality of the theory that emerges from the field of study. The Sherlockian prescription as applied to organizational research is therefore simple, sequential, and reflexive: less theory, better facts; more facts, better theory.

Each case is presented first by a brief vignette providing an exemplar narrative from the case followed by a description of the context of the case. The case is then discussed and related to other relevant cases as well as to other research findings. Van Maanen (1988) characterises narratives as realist, impressionist or confessional. Each of the nine vignettes in the book represents a point of entry into the case and is a realist narrative in the dispassionate third-person passive voice. In order to lend more vitality to the text and to allude to the immediacy of the mobile technology performances studied, the cases are all described and discussed in the present tense. This conveys the feeling of being there while providing detailed insights (Pica, 2006, pp. 84ff). The book emphasises in all cases the experiences of one mobile worker or, in the police case, of two officers in one vehicle. The participants are all anonymous and the names used are fictitious, as are the names of the participating organisations, except London Black Cabs in case 2 and the UK National Health Service (NHS) in case 9. The nine cases will be referred to throughout the book by using the unique identifying number, the participant role and the participant name.

1.4   Book overview

The book is structured into three logical parts across the seven remaining chapters. The first part, comprised of Chapters 2 and 3, explores mobile information technology and mobile technology performances at work, respectively. The second part, Chapters 4, 5 and 6, analyses mobile technology performance based on nine cases. The third part synthesises the findings of the diversity of enterprise mobility portfolios in Chapter 7 and discusses the challenges in managing mobile working in Chapter 8. Figure 1.2 outlines the chapters.

Chapter 2 provides the technological foundation for the analysis by discussing the issues of connectivity, computation, ubiquity and technology intimacy as a pre-cursor for framing the theoretical perspective of services as comprised of affordances as resources for reflexive action and mechanisms scripting actions.

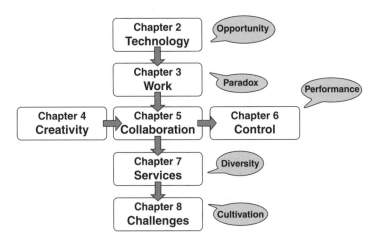

Figure 1.2   The structure of the book with indications of the main theme of each chapter

Chapter 3 reviews mobility and mobile working as creative practices for managing interaction, collaborative practices resolving mutual interdependencies, and controlling practices for planning and management. These practices are situated in the context of the organisational paradox in decisions, for example, between planned interventions and emergent decisions.

Chapter 4 explores creativity in technology performance in individual interventions to cultivate fluid performance and interaction barriers. This is illustrated in cases 1, 2 and 3.

Chapter 5 emphasises horizontal collaboration in terms of mobile technology performances aimed at cultivating transparent negotiation of mutual interdependencies and collaborative barriers. Cases 4, 5, 6 and 7 are introduced here to illustrate this aspect.

Chapter 6 discusses the cultivation of technology performances aimed at the vertical control of mobile working. Two additional cases – 8 and 9 – are introduced, emphasising issues related to the control and management of mobile working.

Chapter 7 synthesises the findings in a discussion of the diversity of enterprise mobility service portfolios.

Chapter 8 concludes the book with a discussion of the challenges of managing enterprise mobility.

Readership

It is my hope that the reader will be able to draw significant lessons about enterprise mobility from this book. It is primarily written as a research monograph aiming at informing the emerging academic discussion of enterprise mobility.

This book can form the basis for graduate-level courses on mobile services in organisations. It can also act as additional reading for graduate and postgraduate students within organisational behaviour, human-computer interaction and work studies. For these groups of students, the book offers complementary reading to the core material, in particular drawing the students' attention to the intricate relationships between the organisation of information work and the specific technical properties of information services.

Finally, it will inform practitioners who wish to explore the possibilities for their organisations to more substantially engage in enterprise mobility projects. Whilst the book will not provide any simple formulae for success, it will instead establish some of the fundamental questions that each organisation needs to ask itself when engaging in significant projects cultivating enterprise mobility. Even if the language at times may be a little academic and the references may come thick and fast, the essential messages should be both relevant and comprehensible for those with less interest in the indepth academic discussion.

2
Technology – Enabling Capabilities

Understanding mobile information technology in the context of work requires a perspective on technology. Throughout the past decades several governing metaphors have been forwarded that characterise computer technology in general as a tool (Ehn and Kyng, 1985), a medium (Andersen *et al.*, 1993), language actions (Winograd and Flores, 1986), intelligent agency (Maes, 1991 and 1994), infrastructure (Ciborra and Associates, 2000) or embodied interactivity (Dourish, 2001). Each metaphor reflects a particular debate and contemporary use of mobile information technology can be viewed as representing all of these perspectives.

The aim of this chapter is to specifically identify the salient capabilities of mobile information technology when applied in the context of work. Orlikowski and Iacono (2001) call for research theorising the information technology artifact and this chapter discusses technological capabilities as a set of distinct affordances and mechanisms assembled into service portfolios. Orlikowski and Iacono identify five different technology perspectives within IS; 1) the tool view; 2) the proxy view; 3) the ensemble view; 4) the computational view; and 5) the nominal view. The perspective presented in this chapter primarily falls within the tool view in terms of information processing, labour substitution, and a tool for increased productivity and social relations. However, the subsequent chapters investigating the use of mobile information technology in the context of work adopt an ensemble view of technology as 'an evolving system embedded in a complex and dynamic social context' (Orlikowski and Iacono, 2001, p. 126).

This chapter outlines salient aspects of mobile information technology in terms of affordances and mechanisms. Affordances are understood as resources for reflexive action, whereas mechanisms represent scripts stipulating and mediating activities. These two are subsequently synthesised into the notion of mobile services. Affordances and mechanisms are discussed in terms of the following six categories: connectivity, portability, memory, pervasiveness, intimacy and priority.

2.1   Connectivity

The beginning of the twenty-first century signalled the era of expanding network connectivity, with the ubiquitous mobile phone becoming an integral part of life, transforming us all into 'digital flâneurs' (Kopomaa, 2000) wandering the streets engaged in casual conversations with remote others. However, the ability to engage in interaction with information sources and people in remote places is not a new phenomenon. Pheidippides became the most famous messenger of all time when in 490 BC he allegedly ran the 26.2 miles from Marathon to Athens to announce the Athenian defeat of the Persians. A variety of relatively expensive means for transmitting messages subsequently deployed combinations of humans, horses, pigeons, motorcycles and aeroplanes to post messages.[4] The exclusive reliance on the physical transportation of messages was set to change when Jean-Antoine Nollet in 1746 transmitted a current through 200 monks each holding onto a short iron bar, thereby proving that electricity can be transmitted over large distances (Standage, 1998). This discovery subsequently led to the global web of telegraph wires enabling message transmission at the speed of electricity, and as a result the Victorians had to deal with information overload (Standage, 1998). The global telegraph set the stage for the next 150 years of interaction explosion.

From telegraph to mobile

Lars Magnus Ericsson was at the forefront of the mobile revolution when in 1910 he installed a telephone for his wife Hilda's car – even if Hilda needed to stop the car to connect the phone to telephone wires (Agar, 2003, p. 8). Two-way closed radio systems used by the military and the police during the 1930s and 1940s mark the first significant use of mobile information technology. Although in 1954 Harold S. Osborne predicted the mobile phone with video connection and a unique personal phone number following the owner from birth to death (Ling, 2004), the basic mobile phone was not a reality until 3 April 1973. Frequency re-use in cells by then allowed the scaling of wireless communication beyond official institutions and Motorola's Martin Cooper made the first call on a hand-held mobile phone to his rival Dr Joel Engel at Bell Labs. The first mobile phones were bulky and either permanently installed in cars or carried between the car, office and home. They were 'carphones', which were considered as a means of turning the car into an office (Agar, 2003).

During the 1980s first-generation (1G) analogue mobile phone networks, such as NMT, were established in cities around the world servicing smaller and smaller mobile handsets. This led to broader adoption beyond a select class of executives and to a lower cost of ownership. The 1990s brought 2G digital infrastructure and handsets, with the GSM standard covering by far the largest number of users. A growing installed base of millions of mobile phones saturated Europe and South East Asia by the mid-1990s. In 2001 the first 3G network was established in Tokyo and the same year also marked the tipping point of 358 million European mobile phone subscribers superseding the 330 million fixed line subscribers (Castells *et al.*, 2007). In 2010 over five billion mobile phone connections globally almost tripled the two billion Internet connections (Kluth, 2008; ITU, 2009; GSMA Mobile Infolink, 2010). Telecommunication standards, such as WiMAX and Long Term Evolution (LTE), provide fourth-generation (4G) wireless connectivity across large distances (Webb, 2010). The beginning of the twenty-first century also signalled an era of Personal Area Network (PAN) technology, where Bluetooth and Zigbee communication standards offer device interconnectivity within close proximity.

Connectivity

The first salient feature of mobile information technology is *connectivity*. Global telecommunications infrastructures have since the early days of the telegraph provided connectivity. Establishing a connection initially relied on expensive and work-intensive relay systems of telegraph offices or human operator telephone switches, but is now extensively automated. Despite significant telecommunications work needed to deliver this automation, the users are able to manage their own connectivity. Global mobile infrastructures and inexpensive mobile devices provide instant connectivity within arm's reach or walking distance for the majority of the world's population. The general notion of connectivity provides an essential part of everyday organisational life, with activities and resources simultaneously interconnected and highly distributed. It is in particular essential to understand enterprise mobility in terms of connectivity as the mobile phone and other mobile information technologies have amplified, distributed and personalised connectivity to an extent not previously seen.

2.2   Portability

The global diffusion of the mobile phone is an example of the combination of connectivity and the miniaturisation of computational devices. Over the past 50 years computer technology has undergone dramatic changes following Moore's Law, which predicted a doubling of the complexity of integrated circuits every 18 months. This has led to an ever-shrinking computer from the initial phase of large mainframe computers in basements to the personal computer on the desk or laptops to PDAs in hands and pockets. Making computers portable, interconnected and embedded is both a significant technological achievement and offers radical changes in the way in which computation is conceptualised.

Human and mechanical computation

The term 'computer' originates from the job description of people engaged in manual scientific, mathematical or administrative calculations (Grier, 2005). Charles Babbage's Difference and Analytical Engines generalised the operations of earlier calculation machines and applied storage in the form of punch cards.[5]

Rise of the mainframe

Computation and other administrative processes have since the advent of scientific management in the 1850s increasingly relied on a variety of information technologies, such as reprographics, filing systems, typewriters and forms (Yates, 1989). Mills (1951, p. 192) argues that in 1930 around one-third of women in offices operated machines other than typewriters and that in the 1950s at least 80 per cent of office jobs could be mechanised. In 1943 IBM's Chairman Thomas Watson famously stated that 'I think there is a world market for maybe five computers' (Berghel, 1999, p. 11). However, from the late 1940s mainframe electronic computers rapidly replaced mechanical computing, heralding an age of large intricate installations of wires, switches and tubes in heavily protected buildings (Richards and Alderman, 2007). Lyons & Co's LEO computer supporting logistics and other administrative procedures was in 1951 the first business application of an electronic computer (Caminer *et al.*, 1998). The mainframe-era computers shaped the ways in which information was managed in organisations, challenged the role of individuals in the organisational management of information and generally automated organisational transactions. This offered both extensive opportunities to automate work, thereby routinising and deskilling work, as well as allowing a broader range of employees to make critical decisions (Zuboff, 1988).

Computing gets personal

In 1977 Digital Equipment Corporation's co-founder Ken Olsen predicted that 'there is no reason for any individual to have a computer in his home' (Berghel, 1999, p. 13). However, the personal computer (PC) marked a significant change, with tiny boxes at first for hobby enthusiasts but subsequently for business use from the late 1970s onwards (Laing, 2004). The advent of spreadsheet applications, such as VisiCalc, was a significant contributing factor to the widespread organisational use of personal computers as it allowed individuals and groups within organisations to make sense of the growing amounts of data. The Apple Macintosh computer in 1985 provided desktop publishing for the first time through a graphical user interface, mouse, PostScript language for communicating with laser printers and Aldus PageMaker (Laing, 2004, p. 144).

Digital communications technology connects organisational computers in Local Area Networks (LANs), combining fixed-line connections and Wi-Fi (IEEE 802.11.a/b/g/n). The networked personal computer is integral to a large proportion of jobs in modern organisations.

Networking computers

The Internet has emerged as a network of networks in developments with origins back to the late 1950s, and a variety of Wide Area Network (WAN) technologies has linked organisational computing across large distances for decades, but the advent of the World Wide Web in the 1990s fuelled the explosive growth in global networking. The combination of a variety of wireless communication technologies and the miniaturisation of computing has moved technology from the desk and closer to the body. In hindsight Bill Gates' statement from the 1980s that 'Microsoft was founded with a vision of a computer on every desk, and in every home'[6] no longer seems particularly visionary when there is quite literally a computer in every person's pocket or handbag. Since the 1980s computer technology has been integrated into a wide variety of household appliances, music players, photographic cameras, etc., providing experiential computing (Yoo, 2010). Furthermore, there has been an explosion in machine-to-machine (M2M) communication where, for example, sensor and computer technologies interact to automate surveillance functions (Lawton, 2004).

Portability

Portability of the computer marks an essential characteristic of the miniaturisation of computing technology. The term 'mobile' in mobile information technology essentially refers to the client part of the technology being portable and therefore implicitly supports users who are geographically mobile. Although the client does not necessarily connect to a wireless infrastructure, this will often be the case. Mobile computing is therefore partly characterised by the combination of *portability* and *connectivity*. The traditional PDA without connectivity is an example of a portable technology that is not mobile.

2.3   Memory

The Turing machine formulated by the British mathematician Alan Turing in 1931 is the best-known abstract representation of the essence of computation (Bolter, 1982). It consists of: 1) an infinitely long tape divided into squares onto which symbols can be printed; 2) a head that can read or write a symbol onto a square; 3) a state register storing the state the machine is currently in; and 4) a table determining what the machine needs to do next – given information of the machine's current state and the symbol under the head. As the tape is infinite, the Turing machine has infinite memory. The machine operates by moving its head forwards and backwards over the tape. It applies the rules given in the table in accordance with the input provided and it will write the corresponding output. Although abstract and conceived long before the first electronic computer, Turing's construct is a precise representation of the strengths and limitations of computational processes. At the centre of this process is the algorithm, the script determining the appropriate symbol to write on the tape given the input read. Algorithms can generally be characterised by combinations of sequencing, selection and iteration (Dijkstra, 1968).

Although memory represents one of the core elements in Turing's mathematical model of computation, it is the understanding of algorithms as unbroken units of input-processing-output that has informed much of the theoretical foundation within computer science (Wegner, 1997). The primary role of Turing's infinite tape can be seen as short-term memory to ensure the unbroken transaction from input to processed output. The data itself represents the forgotten and invisible part of traditional computation (Wegner, 1997). This transaction-based perspective suited the mathematical discourse prior to the invention of the electronic computer, the establishment of computer science as the study of algorithms and the mainframe era of large computer machinery, which chiefly aimed at automating transactional processes exactly.

Interactivity

With the advent of the personal computer and a range of associated technologies, computer technology was applied to a variety of problems beyond those best characterised as discrete transactions, and interactivity emerged as a new perspective on computation (Goldin *et al.*, 2006). The Turing machine and algorithms require that all inputs are specified before computation begins, as would be the case during the days of loading punch cards into a reader and pushing the button. However, this requirement does not serve well as an explanatory model for contemporary use of computer technology where the users and computers engage in ongoing interactive relationships cultivating data (Goldin *et al.*, 2006, pp. 25ff). Interactive computing is proposed as a more powerful means of conceptualising computation (Wegner, 1997; Goldin *et al.*, 2006). Whereas transactions can be modelled as algorithmic processing, objects uniting data and processes can model interactivity. The difference between algorithms and objects can be illustrated in terms of the differences between a sales contract and a marriage: 'Algorithms are "sales contracts" delivering an output in exchange for an input, while objects are ongoing "marriage contracts". An object's contract with its clients specifies its behavior for all contingencies of interaction (in sickness and in health) over the lifetime of the object (till death do us part)' (Wegner, 1997, pp. 81–2). Here *memory* takes on a deeper meaning. It is not merely a necessary aid to ensure that the provided input is calculated into the appropriate output – it is a resource to be reckoned with in its own right.

Memory

The distinction between algorithms and objects is essential in order to characterise the salient aspects of computation as *memory* here takes on a new meaning. This characterisation supports a fundamental distinction between technology that merely assumes interactivity with the user as a series of unrelated encounters and that of user interaction as an ongoing relationship recording and updating data representing aspects of this relationship. Encounters implement the assumption of atomic, unrelated interaction instances. Relationships implement memory of aspects of the interaction (Mathiassen and Sørensen, 2008).

As an example, consider a standard mobile phone, offering the possibility of establishing connections with other mobile phones. Each such connection is treated by the phone as an instance of an encounter between the two phones. No specific assumptions are made regarding the relationships between individual connections other than their chronological ordering and their type – incoming or outgoing. Conversely to this approach, call logs can be organised by person. This is a common approach in contemporary smartphones and to the user represents interactions as a series of ongoing relationships. This organisation of messages does not fundamentally alter the core mobile phone feature of providing possibilities for encounters. It offers the additional feature of these encounters being organised as a series of relationships. By assuming that encounters are indeed part of ongoing interaction relationships, the technology can provide support for the management of these relationships.

The distinctions between algorithms and objects, and transactions and interactions, essentially point toward distinctions applied by a number of researchers: for example, the categorisation of encounter and relationship services (Gutek, 1995); discrete transactions and the continuous delivery of services (Lovelock, 1983); transactional and relational marketing practices (Coviello and Brodie, 2001); and transaction and relationship economics (Zuboff and Maxmin, 2002). The shift from computation primarily being concerned with the transactional automation of back-office processes towards a diverse range of other usages places interactivity, memory and relationships at the centre of our understanding of contemporary computing in general and of mobile computing in particular.

2.4 Pervasiveness

In 1968 Alan Kay and others from Xerox Parc formulated the conceptual design of the Dynabook, which in essence was a notebook or tablet computer. This represented a leap in the understanding of computational support for human activities. In the early 1990s his colleague Mark Weiser (1991) formulated the even more radical vision of the integration of computers into everyday life. Weiser predicted that each person would interact with hundreds or thousands of computers rather than only one. Computers would be embedded in the environment or in portable objects. They would provide ubiquitous environments for engaging with computational objects as easily as with other everyday objects. Weiser's proposition goes beyond the portability of computers and formulates the vision of ubiquitous computing. It signals the disappearance of the computer as a distinct entity.

A variety of research efforts have been concerned with ubiquitous computing. Research programmes such as the Disappearing Computer (www.disappearing-computer.net), research conferences (for example, Ubicomp) and academic journals such as the *Journal of Personal and Ubiquitous Computing* have been established to focus research on this subject. From a technical perspective, ubiquitous computing can be characterised in terms of addressing the following three concerns: natural interfaces; context-aware applications; and automated capture and access (Abowd and Mynatt, 2000). Some research is concerned with the core issues of design and usability (for example, Greenfield, 2006), whereas other strands consider broader relationships between people, ubiquitous computing and the built environment (for example, McCollough, 2004; Bassoli *et al.*, 2007).

While much attention in the HCI field has been devoted to understanding relationships between individual users and the technological interface, little attention has been devoted to the general understanding of unfolding real-life user experiences with ubiquitous computing technology (Abowd and Mynatt, 2000; Sørensen and Gibson, 2008). Within the IS field, very little effort has been invested in understanding ubiquitous computing at all (Yoo, 2010).

Portability and pervasiveness

The term 'ubiquitous computing' is often used as a signifier of the omnipresence of technology (see, for example, McCollough, 2004). Lyytinen and Yoo characterise ubiquitous technology as the combination of *portability* and *pervasiveness*. Pervasiveness is defined as the computer's capability of relating to its environment (Lyytinen and Yoo, 2002a). The PDA is an example of a portable technology that is not pervasive as it generally does not relate directly to its environment. A built-in sensor measuring the water level in a tank and sending an SMS message to a server when the level drops below a certain point is an example of a stationary pervasive technology. Figure 2.1 illustrates the distinctions between portability and embeddedness in terms of stationary, mobile, pervasive and ubiquitous computing.

Technology arrangements often combine separate portable and pervasive elements, where one part is embedded into the environment and the other

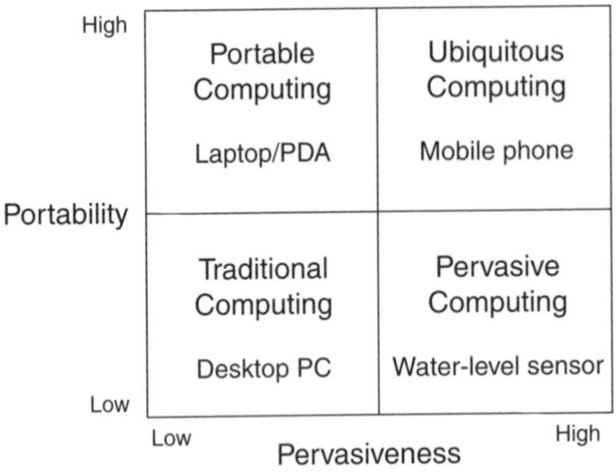

High	Portable Computing Laptop/PDA	Ubiquitous Computing Mobile phone
Portability		
	Traditional Computing Desktop PC	Pervasive Computing Water-level sensor
Low		
	Low Pervasiveness High	

Figure 2.1 Computational portability and pervasiveness. Adapted from Lyytinen and Yoo, 2002a

is portable. Mobile phone technology, for example, relies on the combination of an infrastructure of pre-defined cells and portable handsets moving across these cells. The standard mobile phone has traditionally not provided the user with ubiquitous services, for example, configuring services according to specific aspects of the environment. While the mobile phone at any time is aware of its exact location, the user has traditionally not been informed of this location. Contemporary mobile location-based services do, however, offer the user ubiquitous services and will often explicitly ask the user to confirm the use and transfer of location in applications.

Expanding the perspective of mobile information technology beyond hand-held devices interacting with telecommunication networks offers a range of interesting opportunities for creating experiences by allowing the individual devices and the environment to interact (McCollough, 2004). This merger between the environment and client devices has been characterised as augmented reality and originally emerged in the early 1980s experiments with wearable computing (Barfield and Caudell, 2001).

Pervasiveness

Ubiquity is here characterised as the combination of portability and pervasiveness – the ability of the technology to sense the environment (Lyytinen, 2003) and therefore to be aware of specific aspects of its surroundings. This is not awareness in the traditional sense of the mutual awareness of collaborators (Schmidt, 2002), but rather the systemic awareness stemming from a specific model within the system populated with data from automatic, semi-automatic or manual sensing. One research community seeks to place pervasiveness at the centre, encompassing, for example, ubiquity, mobility, contextual awareness and interactivity (Kourouthanassis *et al.*, 2010). Although complementary, this perspective has not been adopted here as it introduces orthogonal distinctions, for example, conflating portability and connectivity.

2.5   Intimacy

The everyday use of mobile information technology forges close user-technology relationships. Support follows the individual instead of being defined by fixed workstations. Chipchase *et al.* (2004) argue that the belongings people carry with them are associated with fundamental needs. House keys are associated with shelter, money and credit cards with sustenance, and the mobile phone with connections to others and thereby safety.

The body and technology

The intimacy between the user and his or her ubiquitous technology raises the issue of the relationship between the human body and technology, as discussed in several strands of research. This has, for example, theoretically explored the body-machine relationship (Morus, 2002) and the body and mobile information technology (Fortunati *et al.*, 2003). Research has studied the possibilities for humans physically merging with a variety of computer technologies (Mann and Niedzviecki, 2002; Warwick, 2002) and the embodied interaction shortening the distance from intention to action (Dourish, 2001). The relationships between emotions and technology have also been studied in a range of research efforts (for example, Picard, 1997; Fogg and Eckles, 2007; Vincent and Fortunati, 2009).

Emotional and paradoxical interactions

Weiser's vision of ubiquitous computing marks a dramatic departure from the predictions of Watson, Olson and Gates, and has formed a powerful force shaping our understanding of technical possibilities. However, it also engenders notions of ease of use, harmony and computation merging into the fabric of life. Through a much closer technological proximity to the body, the shortening of distance between intention and action, and the ability to connect with remote others, mobile information technology engenders strong emotions with the user. However, this constitutive entanglement (Orlikowski, 2007, p. 1437) is also best conceived as a complex, paradoxical and conflicting relationship and not one characterised exclusively by harmony (Mick and Fournier, 1998; Arnold, 2003).

The individual user's ongoing relationship with ubiquitous computing technology can concurrently consist of elements of control, harmony, interruptions and overload (Arnold, 2003). This can be the case even if the user-technology relationship is entirely reflexive, but is even more so when other individuals and interactive systems actively participate in shaping the user-technology relationship. Arnold *et al.* (2008, p. 49) argue that 'things do not speak': for example, the phone easily disappears from our attention when we recall having spoken with a friend on the telephone – we recall this as having spoken to the friend and not having spoken to the friend through the phone. While this may be true in many cases, things can interrupt the process and demand our attention. This interruption can later be associated with desired interaction, but is nonetheless the technology causing an interruption. Combining intimate technology practices with remote connections to individuals, groups or interactive systems can result in the collective shaping of individual practices.

The rhythms of coupling and uncoupling technology from social action can vary across individuals, situations, tasks and over time (Dourish, 2001, pp. 138ff). The user experience of intimacy is therefore a complex and emerging property of the situation and is shaped by a variety of factors, ranging from practical matters of technology (Sørensen and Gibson, 2008) and situation (Ljungberg and Sørensen, 2000) to fundamental issues of the person's emotional state (Ciborra, 2006). The relationship between a user and his or her mobile phone can turn into an emotionally loaded love-hate relationship where the joys of being able to immediately get in touch with loved ones are interspersed with emails and SMS messages, enabling demands to be placed upon the user irrespective of the situation in which he or she may find himself or herself.

Privacy

Intimate user-technology relationships raise serious issues of privacy and surveillance as, for example, embedding tracking devices support detailed inspections of human movement. Registered contactless payment cards and mobile phones have already provided opportunities for tracking either with proper legal authority or indeed without, and contemporary smartphones embed a variety of sophisticated means for determining user location.

Although privacy and surveillance concerns are key to the broader understanding of mobile and ubiquitous computing, these subjects will not form a core agenda of this book. Such concerns are discussed by a number of researchers: for example, the challenges of identity and the surveillance of citizens (Lyon, 2009; Whitley and Hosein, 2010), and the commercial surveillance of consumers (Albrecht and McIntyre, 2006). This book does not focus on aspects of privacy or surveillance but rather on the underlying technological properties of embedding priorities.

Intimacy

Intimacy is here highlighted as an essential novel aspect of mobile information technology and of importance for understanding enterprise mobility. It relates to several issues – for example, the close proximity to the human body of the client parts of the technology and the ongoing technology relationship fostered by the user. When combined with *connectivity*, intimacy can relate to the identifiability of the user through the close physical proximity to technology. This is of course one of the significant aspects of the mobile phone, where in the Western world it is predominantly an individual technology that is intimately linked to an individual. Intimacy can also be strengthened by the configurability of the technology (Wiredu, 2005): for example, by physical customisation through decoration with stickers and charms (Hjorth, 2009), by customising phone screen backgrounds and ringtones or by customising the suite of mobile applications.

2.6 Priority

Makimoto and Manner's (1997) imagery of digital nomads engaging in friction-free fluid interaction is an easy straw man to critique. However, it is not the simplistic assumptions of the exclusive positive effects of a fluid life that are essential here. More significantly, it is the underlying assumptions of symmetry in social relations mediated by services affording symmetry. Assumptions of fluidity and the absence of friction lead to the assumption of the absence of priorities, preferences, needs, desires, etc. (Ljungberg and Sørensen, 2000; Kakihara *et al.*, 2005). Technology-optimistic accounts emphasise opportunities over realities as the former can be presented as flawless, whereas the latter will be contaminated by real-life messiness.

Symmetry and asymmetry

Mathematically, symmetry is inherently associated with beauty and invariability (Stewart, 2007). Symmetry is geometrically understood as well proportioned and symmetry in a poem makes it pleasing to the ear. It is an ideal that is striven for. This probably conditions us to favour discourses presenting beauty and invariability. However, as argued by Close (2001), asymmetry is not only the creator of the universe as the tiny residual difference at the dawn of time between matter and anti-matter; it is also hidden in all of us, for example, our faces, which look strange if we force symmetry with a mirror. Whilst symmetry is nice, tidy and orderly, asymmetry provides energy, edge and life. Symmetry and asymmetry are concepts associated with a vast array of phenomena, both theoretical and pragmatic, from a variety of academic traditions, such as physics, mathematics, philosophy, genetics, economics and information sciences.

Cyborgs

The imagery of the cyborg presents a perfect symmetric merger of the human body and information technology. High-tech scientists clad in wearable technologies, or having inserted the technology within their bodies, generate media interest and feed into the collective imagination of strange and radical futures.

Since the late 1970s Steve Mann has been experimenting with the long-term use of wearable computing and advocates using such technology for counter-surveillance (Mann and Niedzviecki, 2002). Warwick (2002) takes the concept of embodied interaction literally and experiments with RFID chips and sensors operated within his body, for example, enabling the remote control of a mechanical hand across the Atlantic from sensors inserted into his own hand.

However, with video conferencing, video recording, GPS navigation and the uploading of geo-tagged photos to websites, the mobile phone is emulating much of the same features as Steve Mann has experimented with for years. Mobile phone users have willingly turned themselves into cyborgs with little media storm. The speed by which on 15 January 2009 Twitter user 'jkrums' made his photo of an American Airlines aeroplane landing in the Hudson River[7] available to a global audience while passengers were being evacuated is a testament to mobile phone users globally becoming 'inforgs' (Floridi, 2007).

In the context of this book, the terms 'symmetry' and 'asymmetry' will be used both to characterise technological opportunities and realities of performance. This chapter discusses technology symmetry and asymmetry – technological affordances and constraints. Chapter 3 highlights these in action in terms of symmetry and asymmetry in technology performance and here demonstrates the inherent social asymmetry into which technology symmetry and asymmetry is thrown.

Technology symmetry and asymmetry

Characterising the technological ability to prioritise interaction forms the core aspect of the distinction between technological symmetry and asymmetry. The simple technological standardisation of a connection encounter, i.e., a networking service (Mathiassen and Sørensen, 2008), implements technology symmetry. The technology does not embed assumptions about the ongoing flow of interaction through the standardised connection. Technically it offers an open channel for interaction. Conversely, a mechanism implementing a priority filter for incoming interaction is an example of technology asymmetry. The mechanism either inclusively selects to accept requests, messages or interaction that meet specific criteria, or exclusively rejects those meeting the defined criteria. A co-ordination mechanism (Schmidt and Simone, 1996) embeds a model of how interaction ideally should unfold. This is an implementation of technology asymmetry. The mechanism can, for example, stipulate an ideal process flow between a group of collaborators, each of which update a shared document (Carstensen and Sørensen, 1996).

A standard SMS message sent by one user to another is an example of technology symmetry. Instant messaging systems typically deploy technology asymmetry by providing rules regulating the inclusion or exclusion in spheres of interaction (buddy lists) and filters allowing each user setting its interactional preferences. Technology symmetry leaves the prioritisation to the user. Technology asymmetry supports the user in prioritising interaction, for example, to manage interruptions.

Within the field of information filtering and retrieval, a series of techniques support sophisticated types of filtering, for example, in adaptive filters that learn based on previous occurrences or collaborative filtering deploying mutual recommender functionality (Oard, 1997).

Priority

Technological combinations of *portability*, *connectivity*, *pervasiveness* and *intimacy* raise the issue of *priority*. Technology can provide support for prioritisation. Technology symmetry offers all parties equal status. Technology asymmetry offers possibilities for prioritising interaction through rules stipulating how it should unfold.

Support for prioritisation is relevant for the understanding of intimate human-technology relationships, in particular in the context of work. Organisations operate using complex instruments regulating interactional behaviour, such as standard operating procedures, rules, forms and schedules (Carstensen and

Sørensen, 1996). These are all examples of varying degrees of formalisation of prioritisation.

2.7   Portfolios

The combined forces of the large-scale adoption of digitally converged mobile information technology provide new challenges and opportunities for the development of infrastructures and services (Lyytinen and Yoo, 2002b). This section relates the emerging challenges of digital convergence to the discussion of portfolios as converged bundles of capabilities.

Digitalisation

The core property of digital computer technology, namely the conversion of analogue signals into a digital form leading to streams of binary digits (bits), in principle implies that all signals can be processed by the same component. Digitalisation implies that previously separate technologies can merge and create new ones, and that output from one can become input for another. The properties of the simple line-switched telephone offering connectivity can now be combined with those of other technologies, for example, allowing voicemail messages to automatically be attached to emails and sent to the missing recipient. The combination of *connectivity, portability, pervasiveness, memory* and *priority* can yield interesting portfolios of technological capabilities with digitised inputs and outputs being dynamically re-configured and re-routed between the various elements.

This uncoupling of the previously tight couplings of information types to specific transmission technologies, storage formats and processing technologies implies the emergence of a vast array of possibilities (Tilson *et al.*, 2010). The traditional analogue telephone system implied tight coupling of the handset, which often had to be approved by the national operator, and the analogue transmission format of voice calls defined by international standards. The technical processes of digitising relates directly to socio-technical processes of digitalisation through shared social connections (Tilson *et al.*, 2010). Voice over IP (VoIP) is an example of where analogue voice connections are digitised and transferred from the traditional line-switched telecommunications infrastructure to a packet-switched digital infrastructure.

Data explosion

Computerisation has in the twentieth century led to a rapid growth in the production of data. Extensive digitalisation results in organisational challenges of managing both data and meta-data (*The Economist*, 2010). The data explosion implies that individual organisational members need to engage in complex information management of both data and meta-data. This requires significant technological support enabling individuals to make sense of the data. Rapidly accelerating processes of producing, cultivating, re-packaging and re-appropriating data and meta-data constitute a recursive computational rendition of multiple 'realities' whereby each step leaves more of the analogue to be digitised and thereby subjected to further tagging and re-appropriation. As argued by Kallinikos (2006, p. 157): 'If industrial technology advanced the itemization of the world, computerization is currently involved in its pulverization.'

Portfolios

The mainframe computer represented an attempt to streamline and automate routine processes across the organisation. The personal computer emphasised individuals and groups appropriating the technology to suit specific needs in specific situations. If such specific needs emerge from the situation, then it will be necessary for individual users to be able to appropriate specific functions to these specific needs. The spreadsheet application, for example, functions as an open-ended platform for individual appropriation using a portfolio of possible functions. This implies that the technology, in addition to being considered as an integral part of a streamlined organisational system, can also be viewed as a tool with a complex and individualised relationship forged between the tool and the person using it for his or her purposes (Ehn, 1988). Different individuals can choose different combinations of elements from a given portfolio to reach similar results and such redundancy of capabilities is a characteristic of service portfolios. A portfolio can thereby be likened to a toolbox rather than a streamlined set of procedures, each with specific purposes and with specific rules for how they are used in combination with other procedures. Individual users can both configure their own particular portfolio as well as develop individual practices in how the portfolio is brought into use. It is a configural rather than a standard technology (Burton-Jones and Gallivan, 2007).

2.8   Affordances

Throughout this chapter a number of technological characteristics have been presented: *connectivity*, *portability*, *memory*, *priority*, *pervasiveness* and *intimacy*. Enterprise mobility is here founded on the assumption of heterogeneous adoption and use patterns of portfolios of these capabilities. The following two sections discuss these capabilities in terms of affordances and mechanisms, which are subsequently synthesised into the notion of mobile service portfolios.

Defining affordances

The concept of technology affordances supports the untangling of interrelationships between technological opportunities and social practices applying technology. Such a distinction assumes the possibility of a complex relationship between the intended use as envisioned by the designer and the actual outcome when the technology is in the hands of the end-user. Figure 2.2 illustrates this distinction and also highlights the relationship between a physical artifact and informational objects. The distinction between the intended and the actual use can be characterised as technology affordances and performances (Arnold, 2003). Technology affordances signify opportunities for action, whereas the performances are the actual unfolding actions evoking certain affordances for specific purposes.

Gibson's (1977 and 1979) original definition of affordances denotes an object's latent action possibilities independent of perception. It was aimed at characterising object-action interrelationships as part of perception-action loops (Turner, 2005). Norman (1988) popularised the concept in his discussion of the design of everyday objects and it has since been applied to a range of discussions beyond Gibson's original aims (Turner, 2005, p. 798). Whereas Gibson's original definition emphasises the possible usefulness of a technology, Norman's perceived affordance emphasises the usability of a technology (Gaver, 1991). This is best illustrated by the example of the design of door handles in such a way that it is immediately perceived from the setup whether to pull or push (Norman, 1999, pp. 87–8). Norman's notion of perceived affordances has been widely used within the HCI literature in discussions of usability, but much less so within IS or other fields concerned with the use of information technology. However, within HCI, the term has often been used in a simplistic manner associated with assumptions of a simple linear relationship between specific design choices and usability (Norman, 1999).

Complexity, for example, stems from the problems of applying the concepts of affordances and constraints in the discussion of information technology design (Norman, 1999). Simply physical objects such as door handles can be designed to fully reveal affordances and constraints, and thereby make them more usable. Information technology signifies a complex relationship between perceived,

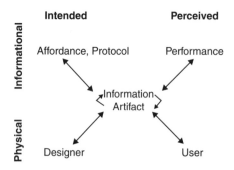

Figure 2.2   Intended affordances and protocols versus performances (Mathiassen, 2010)

physical and hidden affordances (Gaver, 1991; Norman, 1999). The designer may have intended the user to perceive certain performances and thereby hopefully will have produced a usable system. However, users may not perceive all affordances as available to them, as some may be hidden. At the same time, users may mistakenly perceive false affordances that indeed are not available. Issues of technology usability are not at the centre of the discussion in this book. Rather, mobile information technology is primarily characterised in terms of latent possibilities for future action. This matches the distinction between the usability and utility of an object (McGrenere and Ho, 2000).

The notions of affordances and constraints do not characterise essential aspects of the technology, but rather embed assumptions relating to the socio-technical entanglement. Perceived affordances are not inherent properties of the technology but are ultimately contextually dependent upon the perceiver, the situation they may find themselves in and their cultural conventions (Gaver, 1991; Norman, 1999). As highlighted by Dourish (2001, p. 118), affordances are three-way relationships between users, the environments they engage with and the activities they wish to engage in. An ordinary chair only affords a person of average height to sit on it when suitable gravity sticks the person and the chair together.

Affordance research

The term 'affordance' has been applied in a variety of academic discourses and is most frequently related to Gibson's original work, for example, within his field of ecological psychology (Greeno, 1994; Chemero, 2003).

In the context of HCI, Norman's popularisation and re-definition of the term is frequently cited and discussed without an account being given of his notions of constraints or underlying conceptual models (Norman, 1999). A large group of HCI researchers has applied the notion of affordances in discussions relating to the usability of specific innovations. Both Gaver (1992) and Norman (1999) voice criticisms concerning this. The concept has also been applied within cognitive systems engineering (Rasmussen *et al.*, 1994), in sociology to discuss the technological shaping of society (Hutchby, 2001) and within IS, for example, to discuss material agency in flexible working arrangements (Leonardi, 2011) and the practices and affordances of knowledge management (Cook and Brown, 1999). Within the field of computer-supported cooperative work (CSCW), Nardi and Whittaker (2000) propose that the affordances of intimacy, flexible and expressive communication, and visible alerting characterise instant messaging.

There are examples of the concept being applied without explicit definitions or indeed references to key sources, such as Arnold's (2003) discussion of technology or Leonardi and Barley's (2010) review of the literature on the social construction of technology.

Zammuto *et al.* (2007) promote five distinct affordances characterising emerging aspects of information technology in organisations: visualizing entire work processes; real-time/flexible product and service innovation; virtual collaboration; mass collaboration; and simulation/synthetic reality. Zammuto *et al.* (2007, p. 752) consider affordance as a concept bridging information technology and organisational systems through the 'confluence or intertwining of IT and organizational features'. Although Mathiassen and Sørensen (2008) do not use the term 'affordance', their framework can be analysed as promoting the core affordances of organisational information services as portfolios of services.

Enterprise mobility affordances and constraints

The affordances and constraints discussed further in this book relate directly to enterprise mobility and are synthesised from the discussion in the first sections of the chapter with the discussion of affordances in this and the following sections.

The affordance of *connectivity* relates to the discussion of mobile communications. The miniaturisation of computer technology results in the affordance of *portability* and the distinction between stationary and portable aspects of computational provision. Increased interactivity in the use of computers implies the affordance of *memory* where aspects of the ongoing human-computer interaction are recorded and represented. This evokes the distinction between encounters and relationships. *Pervasiveness* signifies the distinction between context-free affordances and those equipped with awareness of the surrounding environment. This also relates to the extent to which the work domain is modelled within the technology. The discussion of close physical proximity between information technology and the human body, as well as the logical proximity of intention and action, results in the affordance of *intimacy*. Intimacy is also afforded through individual configurability of the technology and through identifiability, by which a device identifies its user. The affordance or constraint – depending on the situation – of *priority* signifies choices regarding interaction arrangements. Unprioritised interaction is characterised by symmetry, in that the information technology assumes the equality of all parties. Conversely, the prioritisation of interaction implies the technological mediation of interaction asymmetry. The discussion of digitalisation and convergence leads to the notion of *portfolios* re-combining existing affordances and constraints.

The concept of affordances establishes clarity in the distinction between the possible (design) and the actual (use), thereby replacing the uneasy distinction between and interrelation of social and technical systems (see Figure 2.2). Affordances and constraints merely denote a contextually appropriated diversity of technological opportunities. In relation to enterprise mobility, the notions of affordances and constraints need to be placed in the context of organisational decision-making. This is done in the following section by relating affordances and performances to organisational information services.

2.9 Services

Norman (1999, pp. 41–2) laments the excessive use of screen-based interaction as opposed to more tactile and less ambiguous physical knobs, sliders and buttons constituting less abstract objects and actions. Although affordances can be considered to be aggregated in portfolios as argued above, the notion of an affordance implies that these are resources for reflexive action by the user more akin to the paradigm of direct manipulation than to that of semi-autonomous agency (Schneiderman and Maes, 1997). However, the organisational use of information technology consists of a range of mechanisms that cannot easily be explained entirely as resources for reflexive actions through direct manipulation.[8]

Technological mechanisms are in organisational contexts not only open-ended resources for reflexive action but can also take on the role of formal constructs – scripts. They can be aggregated into more formal organisational objects, acting as cognitive maps informing the user of possible actions, as well as scripts stipulating specific courses of action (Schmidt, 1999). A teaching time-table is not merely a technology affording or constraining a specific action in a specific situation; it is an entire programme of action applied by a variety of participants. Assuming these comply, the timetable will support the orchestration of people, rooms and timeslots according to the master plan. In the context of work it is therefore important to understand for whom the affordances and constraints are provided and the extent to which they are considered as either programmes for action, scripts or merely resources for reflexive action.

Affordances and mechanisms

Whereas Norman's physical buttons and sliders offer good metaphors for affordances as resources for reflexive action, the script offers a similar metaphor for mechanisms stipulating or mediating actions. A script or protocol implies the design-embedded desire to bring about a specific outcome and not merely the provision of a resource. Schmidt and Simone (1996) formulate the notion of the co-ordination mechanism supporting the co-ordination of collaborative activities. They define a mechanism as a construct combining a protocol (or script) of action and an artifact – a permanent construct objectifying the protocol (pp. 165ff). In the context of co-ordinating organisational collaboration, the protocol serves the purpose of a pre-defined organisational construct mediating or stipulating action (p. 166).

Schmidt and Simone's (1996) definition of the computational co-ordination mechanism will be adopted here, although the application of computational mechanisms in mobile working will serve broader purposes than collaborators' co-ordinating activities. Individual mobile workers can, for example, establish scripts automating aspects of their individual information management. Furthermore, the distinction between an individual worker engaging in sense-making through a variety of technologies and the role of the same technologies as support for the worker's engagement in collaboration is analytical, not essential (Schmidt, 1993). A schedule, for example, serves both the purpose of arranging distributed behaviour and as a cognitive aid for the individual acts.

The protocol stipulating activities can be manifested in a number of different artifacts, for example, a timetable printed on a piece of paper or computerised as a set of calendar alarms on a mobile phone. Different artifacts embedding such a timetable protocol can display a variety of affordances. If a computerised timetable offers interactivity, then this could, for example, support an ongoing relationship allowing pupils to keep a record of their own attendance.

Maps or scripts?

The extent to which social action can be stipulated and determined by a predetermined protocol is the subject of significant debate far beyond the scope of this book – see, for example, Schmidt and Simone's (1996, pp. 165ff) discussion of protocols; Schmidt's (1999) distinction between maps and scripts; Suchmann's (1987, 1994 and 2006) persistent defence of situated action; and the associated debate in the *CSCW Journal* by a range of academics, including Winograd (1994, pp. 193–4), arguing that increased complexity implies the need to stipulate the organisation of activities – an argument also forwarded by Carstensen and Sørensen (1996). Within the context of this book, the argument is both interesting and relevant but also slightly tangential, as the concern here is not to settle to what extent a protocol can be seen as a map – a resource for situated action – and to what extent it is a script determining a course of action. As with the discussion of affordances in terms of embedding assumptions relating to context and user, mechanisms will embed assumptions of context and proper use. The primary concern here is to investigate the diversity of services (affordances and mechanisms) in enterprise mobility. Schmidt's (1999) analysis of maps and scripts provide the conclusion that although the specific role of formal constructs at work is under-researched, it is not reasonable to assume that these not only take up the role of maps but can also serve as scripts, reducing overheads through stipulating action. However, the distinction between maps and scripts still maintains the notion of resources for situated action as exceptions can cause human agency to deviate from both maps and scripts (Schmidt, 1999).

Services

Having defined information services in terms of affordances and mechanisms, this section takes a closer look at how information services can be characterised within the context of organisational activities. Jointly, affordances and mechanisms provide portfolios of organisational information services.

Information services form the link between business and software services (Mathiassen and Sørensen, 2008, p. 314). Organisational information services are affordances and mechanisms implemented through software services, which in turn are provided through computational processes. Software services represent the basic operations of drawing graphical widgets on screens, retrieving data from the database and performing basic calculations. Business services are provided through business processes enabled by organisational information services. Business services relate to the organisational domain of activities and not to the specific provision of technological support to deliver these services. As an example, the business service of managing investments within the financial sector can be supported by a range of business processes related to the management of accounts, transactions and investment options. These processes can in turn be supported by a range of information services supporting decisions, collaboration and planning.

Information services can, as discussed previously, implement memory of the ongoing interaction, assumptions of symmetry or asymmetry of the interaction, etc. The aim of emphasising information services as the overall unit of analysis for enterprise mobility support is the emphasis on the abstract notion of support for decisions and information management rather than the specific artifactual embodiment of this support. Figure 2.3 highlights the focal enterprise mobility services in terms of the following six categories: connectivity, portability, memory, pervasiveness, intimacy and priority.

2.10 Summary

The properties of mobile information technology can be characterised as mobile service portfolios comprised of affordances and mechanisms. Affordances signify

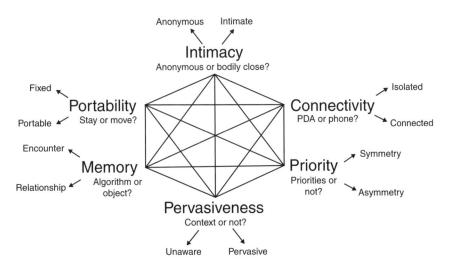

Figure 2.3 Overview of enterprise mobility services

resources for reflexive action. Mechanisms contain protocols aimed at scripting user behaviour. Mobile service portfolios combine six categories of services: 1) *connectivity* with others or with remote information services versus purposeful isolation; 2) *portability* of devices and services; 3) *memory* of ongoing relationships as opposed to a series of isolated encounters; 4) *pervasiveness* recording aspects of the service environment; 5) *intimacy* with users in terms of possibilities for individualisation closely associated with and identifying the user; and 6) *priority* as services supporting the stipulation of technology asymmetry. The following chapter explores mobile work as technology performances evoking these services.

3
Work – Facing Paradoxes

This chapter presents the foundation for understanding mobile work in terms of mobility, mobile working and mobile interaction. It then introduces the notion of organisational paradoxes as a way of understanding the conflicting organisational requirements and tensions mobile workers engage with in their daily work. The notions of organisational paradox and tension are then related to planned and emerging technology performances. Technology performances are then analytically separated into three types, each emphasising a particular perspective on mobile work: 1) creativity, which denotes the mobile worker engaging in managing interaction; 2) collaboration, where the mobile worker negotiates mutual interdependencies with others; and 3) control, signifying activities aimed at overseeing, planning and managing mobile work. Each of these three categories will be discussed in Chapters 4, 5 and 6.

3.1 Mobility

The term 'mobility' is more a conceptual clearing than a well-established concept, as there are many ways of using the terms 'mobile' and 'mobility' depending on the perspective applied. The concept of 'mobility' or 'mobile' can generally signify 'the ability and willingness to move or change' (Wikipedia. org). The term 'mobility' is used to characterise a range of phenomena in a variety of contexts, such as: electron mobility from solid-state physics; the mobilisation of armed forces; economic mobility; social mobility of individuals and families relative to social stratification; population mobility in terms of migration; and geographic mobility of professions, tribes and other groups. Mobility as broadly related to the definition and use of physical space relates to research within geography, architecture, sociology, organisational theory and philosophy.

Philosophical discussions explore our general understanding of place and space in contemporary life (Tuan, 1977; Lefebvre, 1991; Casey, 2009). Tuan

(1977) emphasises the interdependency of space and place, and formulates these as unified by common experiences. Geographers have a long-standing interest in how social and economic development transforms our understanding of space – for example, how social space is extended under globalisation (Gregory and Urry, 1985; Massey and Jess, 1995). Research also seeks an indepth understanding of everyday human movements (Hägerstrand, 1975; Carlstein, 1983), changes to everyday travelling (Peters, 2005; Pooley *et al.*, 2005), the relationships between dwelling and moving (Morley, 2000), the diverse range of methods for studying mobilities (Büscher *et al.*, 2011) and mobility as displacement in general (Cresswell, 2006). Nandhakumar (2002), for example, applies the time-geography techniques of social geographers to understand locational and spatial aspects of software development work.

Architects have also sought to understand how to facilitate new ways of social engagement as well as working practices through new workplace design (for example, Laing *et al.*, 1998; Becker, 2004; Myerson and Ross, 2006). The key challenge is one of organising spaces to best support the rhythms of situated interaction through a variety of architectural mechanisms. Physical arrangements will then, for example, be influenced by the type of organisation, the type of work, the extent to which work is individual and the degree of interaction with others required to accomplish the work. The requirements for spatial organisation of work can, for example, vary depending on the degree of autonomy and the need for interaction with others. Laing *et al.* (1998) distinguish between the degree of participant interaction and the degree of autonomy in work, and outline four archetypical configurations of workspaces: the *hive* is suitable for organising individual routine work with a low degree of autonomy and little interpersonal interaction; the *den* is appropriate to house highly interactive groups where there is a low degree of individual autonomy of work; the *cell* is appropriate for work with a high degree of autonomy and little interpersonal interaction; and the *club* is a configuration of space particularly suited for the engagement in knowledge-intensive working where there is a high degree of interpersonal interaction and a high degree of individual discretion. This characterisation highlights the essential role of location as both a facilitator for intense interaction and as a filter for creating peace and quiet. It also highlights the importance of organisational arrangements for facilitating the sharing of knowledge, as discussed by Nonaka and Konno (1998) using the Japanese concept of 'Ba', which is an abstraction beyond mere physical arrangements of interaction in organisational spaces.

3.2 Work

The understanding of mobile work relates directly to insights from research into other flexible working arrangements – for example, flexible working arrangements, home working, distributed work and global teamworking. This section discusses such flexible work arrangements in relation to mobile working.

Flexible work

Within the sociology of work, Mills (1951) offers the analysis of white-collar work with some emphasis on the practical and spatial arrangements of work. Goffman (1959) offers detailed sociological analyses of the use of space as a means of engaging in social interaction and also suggests a spatial dramaturgical metaphor for such interaction denoting the differences between frontstage and backstage interaction. His work has been used in discussions of mobile communication (for example, Ling, 2008). There has been significant discussion among both industry practitioners and academics on the issue of flexible working arrangements and late twentieth-century changes to working life, most famously perhaps the first part of Castells' trilogy formulating the global transformation of institutions, economies and work as spaces of flows (Castells, 1996). Other accounts of contemporary changes to working life explore work and family life (Hochschild, 1997; Carnoy, 2002), the intensification of work (Bunting, 2004) and flexible project working (Kunda, 1992; Clegg *et al.*, 2006).

Although a significant aspect of the changes to modern working life is the issue of flexible working arrangements, human geographical mobility and the role of space in organisational life are largely under-researched issues (Felstead *et al.*, 2005). Furthermore, each discipline has its own perspective on the understanding of the role of space (Dale and Burrell, 2008, p. 5). Dale and Burrell offer a critical analysis of the intertwined physical and symbolic role of space in organisations in terms of the shaping of identity and power. This work is partly based on Lefebvre's (1991) philosophical discussion of the meaning of space, in its broadest sense, as experienced in everyday life.

Distributed work

A number of studies have explored new workplace arrangements (Holman *et al.*, 2003). In terms of characterising the mobility of work and of workers – work outside the traditional workplace – a number of related concepts have been used to characterise deviations in work arrangements from the traditional factory workstation or desk-based office work. These include, for example, SOHO (small office/home office), where the traditional office is split between a smaller office at work and the addition of a home office (Felstead and Jewson, 2000; Felstead *et al.*, 2005, Chapter 5); telework and telecommuting (Jackson and van der Wielen, 1998; Daniels *et al.*, 2001); remote work and distributed work (Sproull and Kiesler, 1993; Hinds and Kiesler, 2002); globally distributed teamworking (Kotlarsky *et al.*, 2008); and work on the move (Felstead *et al.*, 2005, Chapter 6). In particular, telework has been the subject of significant research (Daniels *et al.*, 2001; Lamond *et al.*, 2003). Much of this work characterises the combined working from home and from the office during the week, or perhaps working entirely from home.

Daniels *et al.* (2001) suggest a typology of teleworking in terms of the following work characteristics: knowledge-intensity of the work; intra-organisational contact; extra-organisational contact; information technology usage; and the type of telework arrangement (whether home-based, remote office and nomadic). These dimensions are then applied to characterise a range of organisational jobs, so, for example, an engineer could be engaged in information technology-intensive and knowledge-intensive nomadic work with high degrees of both extra- and intra-organisational contact. At the other end of the spectrum, a proofreader may be engaged in home-based work with a low knowledge intensity and a low degree of extra- and intra-organisational contact, and may not need information technology.

A significant body of research stemming from different fields has studied distributed teams and the role of information technology support for teams, including the following areas: practical advice regarding the organisation of remote work (Nilles, 1998); requirements for establishing collaboration across distance in terms of common ground among collaborators, work-task coupling, motivations to collaborate and technology sophistication (Olson and Olson, 2000); and patterns of collaboration in teams as either managerial-, expertise- or team-centred (Perlow *et al.*, 2004).

Much effort has been devoted to the study of global virtual teams consisting of group collaboration across time zones – for example, the opportunities and challenges of knowledge work across globally distributed teams (Kotlarsky *et al.*, 2008); the creation and maintenance of trust in globally distributed virtual teams (Jarvenpaa and Leidner, 1999); team-member awareness of collaboration (Leinonen *et al.*, 2005); and temporal aspects of global trading (Barrett and Scott, 2004). Concerning the distinction between virtual and non-virtual teams, Hughes *et al.* (2001) provide a critical account and question the extent to which the 'virtuality' label adds analytical power in explaining the unfolding of everyday work activities.

A characteristic of the studies of virtual teams is an emphasis on collaboration across remote teams – the mobilisation of work activities between distant teams – with less detailed interest in the mobility of individual team members. Similarly, teleworking research emphasises remote, but stationary, individual workers. Much of the research in remote working and teleworking also emphasises the highly pragmatic challenges of making teleworking work in terms of providing adequate information technology and managerial support. Some research is concerned with understanding the changes from an organisational theory or Human Resource Management (HRM) perspective. This research can contribute to the inquiry in this book, as the core issue is the fundamental understanding of technologically mediated interactions at work in organisational contexts. However, as this research classifies and brackets specific types of work as tel-ework, and thereby introduces a specific category, it also desensitises the analysis of what could be argued as much more fundamental aspects of mobile working. In short, there are many subtle categories of mobility. Furthermore, although Daniels *et al.* (2001), for example, argue that telework can be characterised in terms of home-based, remote office and nomadic, most telework research focuses on people working either from home or from remote offices. The emphasis on human resources issues and on homeworking places the teleworking discussion close to the debate regarding flexible working.

Mobile work

Mobile work is still far from defined as a distinct field of research separate from studies of flexible and distributed working. Hislop and Axtell (2007) call for more research into the spatial mobility of work within organisation studies, and in general there has not been much emphasis on the role of human mobility as a means of co-ordinating work within this field. A few studies have specifically explored mobile working, such as managerial and information work (Felstead *et al.*, 2005, Chapter 6); distributed repair engineers (Orr, 1996), itinerant software work (Barley and Kunda, 2004), medical trial work and information technology support (Ljungberg, 1997), telephone engineers (Wiberg, 2001), equity investors (Mazmanian *et al.*, 2006); the police (Manning, 2003; Sawyer and Tapia, 2006; Straus *et al.*, 2010), cab drivers (Skok and Kobayashi, 2005) and management consultants (Sturdy *et al.*, 2009).

Enterprise mobility

Enterprise mobility denotes the study of mobile information technology in the context of mobile work. As such, it represents a mere analytical categorisation of studies of mobile work, as such activities generally rely on various forms of mobile technologies. However, the categorisation gains validity as many of the studies of flexible and mobile work merely treat technology as an additional aspect as opposed to investigating the mutual constitutive aspects of techno-logical possibilities in work practices.

Considering the challenges posed by Lyytinen and Yoo (2002b) in their outline of an IS research agenda for mobile and ubiquitous computing, most of these challenges are concerned with organisational aspects of mobile infor-mation technology development and use, and a decade later there is still the need for significant research efforts to address all of these. It is, for example, notable how the intimate co-construction of flexible working arrangements and the use of a variety of information technologies both form a foundation for Sturdy *et al.*'s (2009) research, but at the same time is almost non-existent in the discourse and is discussed in terms of boundaries and knowledge flows.

Furthermore, although the transformation of work and flexible working played a significant role in Castells' (1996, pp. 216–354) analysis of the network society, the influence of mobile information technology on work is not given a similar prominent treatment in his book on mobile communication and is only briefly mentioned (Castells *et al.*, 2007, pp. 78–83).

A number of edited collections explore various aspects from an academic perspective of the use of mobile and ubiquitous information technology to support the conducting of work within and between organisations – for example, enterprise mobility (Basole, 2008), mobile working (Andriessen and Vartiainen, 2005; Hislop, 2008), pervasive information systems (Kourouthanassis and Giaglis, 2008), mobile information systems (Krogstie *et al.*, 2005; Andersson *et al.*, 2007) and ubiquitous information environments (Baresi *et al.*, 2004; Sørensen *et al.*, 2005).

There are also very few business books addressing the issues of using mobile information technology to support work – for example, the general prediction of global digital nomads roaming the earth unhindered (Makimoto and Manners, 1997), advice for the implementation of mobile working, using experiences drawn from Nokia's own practices (Lattanzi *et al.*, 2006) and a variety of practitioner and executive how-to guides of key enterprise mobility decisions (such as Hayes and Kuchinskas, 2003; McGuire, 2007; Darden, 2009; and Reid, 2010).

In the reality of enterprises deploying a variety of mobile information technologies in conjunction with flexible and mobile working practices, there will also be a significant need to understand how human and material agency is mutually co-constructed or imbricated (Leonardi, 2011). This subject is the chief concern of this book and the following sections establish the conceptual foundation for understanding mobile working practices.

3.3   Defining mobile work

The definition of mobile work relates to that of workplaces. Felstead *et al.* (2005, Chapter 2) suggest three concepts: *workstations*, which are immediate locations where employment tasks are conducted; *workplaces*, which are defined in terms of buildings and other physical constructions supporting one or more workstations; and *workscapes*, which are understood as the total network of workstations and workplaces occupied by individuals and groups engaged in work. They further characterise clusters of network connections incorporated in various combinations into workscapes, and analyse three such clusters, namely working in the collective office, working at home and working on the move. Whereas some workers will engage one workscape, others will engage in work across several workscapes. Multiple workscapes can be characterised in terms of a variety of dimensions (Felstead *et al.*, 2005, pp. 23–36): opportunities for personalised workspaces; the extent of diverse and multifunctional workstations, opportunities for personalised work times; the degree of reliance on information technology; the extent of informal social interaction; opportunities for visual surveillance; opportunities for the display of status symbols to colleagues; the management of work and non-work relationship boundaries; and the extent of participation in the general corporate aesthetics.

Engagement in plural workscapes necessitates *assembly* in order to co-ordinate the different activities. Assembly relates to concepts of collaboratively meshing activities and results (Schmidt, 1993). *Pathways* signify the potential trajectories in workscape choices and the *routes* are the most routinely chosen directions (Felstead *et al.*, 2005, p. 18). When work engages plural workscapes, there will frequently be a need for *repair*, which responds to exceptions, interruptions, disruptions, etc. (Felstead *et al.*, 2005, p. 19). This relates to issues of interruptions and interaction overload (discussed later), and can also be compared with the notion of coping strategies (Jarvenpaa and Lang, 2005). *Wayfinding* implies the knowledge, skills and attitudes necessary for working across plural workscapes, and this includes devices or machines, systems of social discipline and technologies of the self, i.e., 'ways in which people, more or less consciously and reflexively, mobilize and organise their attitudes, practices and feelings in the course of their everyday lives' (Felstead *et al.*, 2005, p. 116). Working across plural workscapes can facilitate workers occupying *personalised space* where it is difficult to engage direct management control by surveillance and thereby undermine presence and visibility. Personal spaces can be subjected to *stalling*, which is the practice of laying claim to a bounded space (Felstead *et al.*, 2005, p. 22) – for example, commuters purposefully occupying the outer of two seats on the train to ensure more space to work (Goffman, 1971).

A simple way of characterising mobility is through the functional distinction between three modalities of mobility: *travelling*, *visiting* and *wandering* (Kristoffersen and Ljungberg, 2000). This distinguishes between movements from A to B when travelling, the relatively fixed working at a desk when visiting and the localised spatial movement within a limited area when wandering. Kristoffersen and Ljungberg (2000) apply this tripartition to characterise the usability of various mobile information technologies and thus link individual geographical mobility to the use of mobile information technology. They argue that mobile technology such as smartphones and PDAs are useful in all three situations. Portable technologies such as notebook computers are only useful when visiting and travelling (specific conditions permitting), whereas desktop technology is only really useful when visiting.

Luff and Heath (1998) make similar, albeit more succinct, arguments in their distinction of three kinds of mobility in collaboration; *micro-mobility*, *local mobility* and *remote mobility*. Micro-mobility characterises the way in which physical artifacts can be handled and managed at arm's length between people. Local mobility signifies the ways in which people collaborate through moving around their place of work, thereby using information and colleagues available to them where they are. Remote mobility implies the wider mobility of actors distributed across distance and moving between these. Here the application of various artifacts and procedures will often support the co-ordination of remotely mobile activities. Luff and Heath describe how an allocation sheet at a building site enables people in the site office to engage in asynchronous and remote co-ordination of activities out on the building site. Bassoli (2010) offers a more comprehensive discussion of human movement and the use of a variety of ubiquitous information technologies She argues that traditional categorisations of places are insufficient and calls for further exploration of the tensions in the co-construction of everyday places and the ubiquitous technologies inhabiting these places.

The analytical categorisation of work into geographically local and remote activities can illustrate the differences between remote but stationary working and local but mobile working. Figure 3.1 illustrates this distinction by characterising the analytical categories of office working, remote working, local working and mobile working.

What then are the consequences of work being both mobile and geographically distributed? Some commentators argue that digital technologies will spell the 'death of distance' (Cairncross, 1997) with anytime, anywhere interaction replacing interaction between co-present participants wholesale. Others argue that 'distance matters' (Olson and Olson, 2000), that successful remote interaction greatly depends on the extent of common ground among collaborators, the degree of work task coupling, the collaborators' motivations to collaborate and their level of technology sophistication. In this respect, distance represents an additional complexity when engaging in collaboration. Paradoxically, according to Dubé and Robey, 2009: virtual teams function best

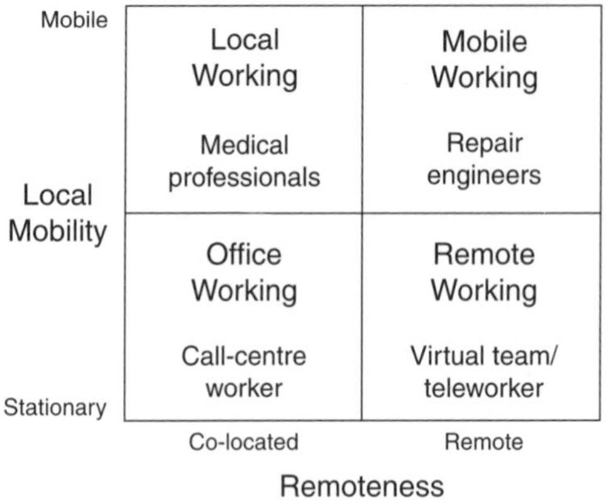

	Co-located	Remote
Mobile	Local Working Medical professionals	Mobile Working Repair engineers
Stationary	Office Working Call-centre worker	Remote Working Virtual team/ teleworker

Local Mobility / Remoteness

Figure 3.1 Local and remote working

when they are at times physically present; structures are critical to enabling flexible teamworking; highly independent contributions make up interdependent teamwork; social interaction is critical for successful task-oriented work; and mistrust is a key condition leading to a sufficiently high degree of team trust. Such paradoxes in enabling effective distributed teamworking require a variety of coping strategies, for example, mixing distributed working with face-to-face meetings and balancing standardised, yet flexible, modes of collaborating.

Fluidity

Social geography research and related philosophical and sociological research on mobility generally discuss the issue of mobility beyond simple conceptualisations of human movement. John Urry (2000), for example, suggests the inclusion of all mobilities in a mobile sociology. According to Urry, we must understand the general mobilisation of society where: '... people travel along transportation scapes for work, education and holidays. Objects that are sent and received by companies and individuals move along postal and other freight systems. Information, messages and images flow along various cables and between satellites. Messages travel along microwave channels from one mobile phone to another' (Urry, 2003, p. 5).

Contemporary life, with its extensive reliance on a variety of technology-mediated interaction, has been described as liquid or fluid in comparison to a relatively stable past. Castells (1996, Chapter 6) characterises the 'social meaning of space and time' as a space of flows as opposed to the traditional space of places in industrial society. Bauman (2000 and 2007) offers a comprehensive insight into 'liquid modernity' with socially re-defined and disappearing boundaries. Mol and Law (1994) propose 'fluid spatiality' as a 'social topology' in addition to the two traditional topologies of regions and networks. The assumption of a region implies delineation through boundaries, and the assumption of networks is the linking of stable relationships. The fluid topology is defined by transformation and variation: 'neither boundaries nor relations mark the difference between one place and another. Instead, sometimes boundaries come and go, allow leakage or disappear altogether, while relations transform themselves without fracture. Sometimes, then, social space behaves like a fluid' (Mol and Law, 1994, p. 643). Urry applies the fluid social topology in his analysis of global fluids: 'The "particles" of people, information, objects, money, images, risks, and networks move within and across diverse regions forming heterogeneous, uneven, unpredictable and often unplanned waves' (2003, p. 60). Dale and Burrell (2008, pp. 117ff) suggest 'liquidity' as a concept characterising contemporary architectural and managerial ideologies of organisational space. Discussions of liquid modernity, flows and fluidity point toward dramatic changes in society. Projecting these changes onto interaction at work emphasises the role of information technology in flexible working arrangements, the social definition of previously given boundaries and the additional activities of negotiating these fluid boundaries (Nardi and Whittaker, 2000; Felstead *et al.*, 2005; Wajcman *et al.*, 2009).

Such a broad and inclusive perspective offers a comprehensive umbrella for understanding a diverse range of research issues related to mobility and is also associated with the risk of in turn treating other important aspects as black boxes. This much broader sociological perspective is founded on the observation that global flows of people, objects, information and images call for deeper inquiry than has been the case. Such a study involves understanding mobilities in terms of: the corporeal travel of people; the physical movement of objects; the imaginative travel of images of places and people; virtual travel transcending geographical and social distance; and communicative travel through person-to-person messages (Urry, 2007, p. 47). The broad social science discussion on mobilities assembles an interesting and rich discourse by growing tentacles that reach deep into a variety of social science issues, as illustrated in Urry (2007). However, not many tentacles reach deep into the discussions of how social affairs and contemporary information technology are mutually co-constructed – constitutively entangled (Orlikowski, 2007) – in everyday life.

Mobile interaction

An essential concern here is to scope the investigation in order to steer clear of the two imminent dangers of Scylla (a sea monster who lived underneath a dangerous rock at one side of the Strait of Messina) and the whirlpool Charybdis, both of which Odysseus was forced to navigate between on his journey.[9] In this study Scylla can symbolise the oversimplified conceptualisation of mobility directly related to human physical movement, whereas Charybdis symbolises an abstract sociological discourse characterising a general mobile sociology. The aim here is to avoid both and subsequently the study is faced with the danger of navigating a difficult strait.

The chosen path through the Strait of Messina is the concept of mobile interaction (Kakihara and Sørensen, 2002). Rather than merely categorising and classifying modalities of human movement in time and space as the core aspect of mobility, further understanding can be gained by the more abstract notion of the mobilisation of interaction through information technology in general and mobile information technology in particular. This is similar to the concept of 'mobile communication' used by several researchers (such as Castells *et al.*, 2007; Kleinman, 2007; Ling, 2008).

One of the consequences of mobile information technology is indeed a dramatic mobilisation of interaction mediated by handsets, tags, readers, services and associated infrastructures. The mobile phone enables personalised, situated and mobilised interaction between people. The technology facilitates new arrangements in the co-evolution of technological opportunities and human practices, for example, in terms of the temporal, spatial and contextual aspects of interaction. Mobile interaction fixes the essential focus of the study as mediated interaction in the context of human activities, mobile or otherwise. Obviously, we can use the concept of mobile interaction to characterise the locationally fixed work of call-centre staff as well as that of highly geographically mobile repair engineers (Kakihara and Sørensen, 2004). However, all the fieldwork cases discussed in this book document both physically mobile workers and the extensive use of mobilised interaction, thereby contributing to focusing our inquiry.

Perry *et al.* (2001) identify four facets of mobile interaction: 1) *planful opportunism* of preparing for the unpredictable by bringing along technologies, documents and resources that may turn out to be useful; 2) engaging in the effective use of *dead time* where work normally would not be possible; 3) using the mobile phone as a *device proxy* when necessary resources are not available – for example, calling back to the office in order to have a fax sent to a customer; and 4) *remote awareness monitoring* and access to colleagues where mobile services support the mobile worker in maintaining awareness of the situation back at the office, thereby supporting a sense of community.

Mobile interaction supports a broad range of social purposes from micro-co-ordination (Ling, 2004) to nurturing deep social relationships (Ling, 2008). The instant availability of interaction results in individually cultivated performances (Arnold, 2003) with emerging rhythms of device use (Green, 2002; Sørensen and Pica, 2005). New interactional repertoires used to establish and maintain social relations evolve as a result of mobile interaction (Licoppe, 2004), as well as new co-present practices (Weilenmann and Larsson, 2001; Ling, 2008). Mobile interaction challenges existing interaction rituals and can both strengthen social relations and challenge co-present intimacy (Ling, 2008), for example, expressed as the absent presence (Gergen, 2002) when interrupting phone calls or SMS messages enters into intimate situations.

The analysis presented in this book emphasises interaction and interactivity as means of engaging in decision-making, as modes of operations with information technology and as the underlying unit of analysis when studying primary phenomena. The following section explores research on organisational paradoxes and tensions as a means of explaining the challenges of enterprise mobility. This will be related to the inherently contrary technology performances of organisational members applying mobile information technology.

3.4   Paradox

Mobile interaction within the context of work will occur in a contradictory and paradoxical environment of conflicting demands, opinions and decisions.

This section explores organisational paradox and in particular the conflicting requirements of members being expected to be contextually ambidextrous. The understanding of organisational complexity and contextual ambidexterity will form a foundation for the subsequent analysis of enterprise mobility.

Organisational paradox

A paradox is 'an apparently unacceptable conclusion derived by apparently acceptable reasoning from apparently acceptable premises' (Clark, 2002, p. 132). However, the paradoxes, tensions and conflicting demands experienced by mobile workers do not comply with this definition entirely. The paradoxes discussed here are distinctly different from philosophical, logical and mathematical paradoxes. One such example is the paradox of Achilles and the tortoise (Clark, 2002, p. 1) where the speedy Achilles will supposedly never catch up with the tortoise, who leaves behind smaller and smaller intervals forever separating the two combatants. As in reality Achilles would catch up with the tortoise, so did the mathematical formalisation of infinite sum limits eventually catch up with this particular mathematical paradox.

Cameron and Quinn (1988) suggest that the exploration of organisational phenomena can be enriched through a theoretical lens acknowledging contradictory but at the same time related elements, which seem paradoxical when considered simultaneously. The essence of organisational paradox is the contradictory situation of opposite solutions both making sense without any straightforward best solutions. The perspective of organisational paradox thereby refrains from forcing conflicting and contradictory findings to emerge from the 'complexity, diversity and ambiguity of organisational life' (Lewis, 2000, p. 760) into oversimplified categorisations.

A firm definition of organisational paradox has yet to emerge (Chae and Bloodgood, 2006, p. 4) and the term has been used in a variety of ways, often as a label characterising conflicting demands or what seem to be illogical findings not revealing much substance (Lewis, 2000, p. 760). Organisational paradoxes are comprised by a wide variety of interrelated elements, such as individual practices, perspectives, feelings and interests (p. 761). These paradoxes are socially constructed as participants polarise complex interrelationships into simple tensions when making sense of complex, dynamic ambiguity. Paradoxes reveal themselves through reflection and interaction, making irrationally co-existing opposites apparent (p. 761). The explanation of organisational phenomena in terms of contradictions can be seen as poor research, but it may indeed forward a more comprehensive understanding of organisations inherently operating under conditions of ambiguity and complexity (Poole and Van de Ven, 1989, p. 562; Lewis, 2000).

The organisational paradox perspective has been applied in the study of a range of issues, such as: organisational change (Ford and Backoff, 1988);

innovation (Singh *et al.*, 2009; Andriopoulos and Lewis, 2010); managerial sense-making (Lüscher and Lewis, 2008); organisational governance (Sundaramurthy and Lewis, 2003); product development (Lewis *et al.*, 2002); managing creatives through identity regulation (Gotsi *et al.*, 2010); paradoxes as invitations to act (Beech *et al.*, 2004); the management of knowledge through information technology (Chae and Bloodgood, 2006); large-scale technology management (Zheng *et al.*, forthcoming); the role of information technology in organisations (Robey and Boudreau, 1999); virtual teamworking (Dubé and Robey, 2009); and telecommuting (Pearlson and Saunders, 2001). Paradox thinking has not escaped popular management writing, for example, Handy's (1995) and Naisbitt's (1994) pragmatic accounts of the diversity of paradoxes facing those engaging in modern life or Farson's (1996) exploration of the paradoxes of management.

Pool and Van de Ven (1989) suggest that a paradox perspective can serve as the foundation for developing management theory and promote four approaches to resolving organisational paradoxes to be used individually or in combination: *transform* – accept the paradox and learn to live with it; *eliminate* – resolve the paradox through spatial separation; *avoid* – resolve the paradox by situating at different temporal locations; and *transcend* – resolve the paradox through synthesis. The four categories are here named as in Beech *et al.* (2004).

Contextual ambidexterity

Quinn and Rohrbaugh (1983) explore how effective organisations seek to simultaneously fulfil conflicting goals and suggest a competing values framework – for example, by balancing competition between internal and external focus and between the need for organisational control versus flexibility. A number of researchers have studied the conflicting demand on organisations to concurrently attend to immediate operational issues and strategic opportunities. The 'ambidextrous organisation' is able to concurrently deal with *exploitation*, i.e., immediate operational issues exploiting existing capabilities, and *exploration*, which is the strategic challenge of re-defining essential aspects of the business in order to take account of new opportunities (Birkinshaw and Gibson, 2004; Gibson and Birkinshaw, 2004; O'Reilly and Tushman, 2004; Raisch *et al.*, 2009; Andriopoulos and Lewis, 2010). A similar discussion of survival patterns for software organisations has been characterised in terms of balancing improvement and innovation (Holmberg and Mathiassen, 2001).

Structural ambidexterity denotes the implementation of different structures, each aimed at one particular type of activity. Jointly these structural interventions are intended to render the organisation able to balance conflicting concerns. Conversely, contextual ambidexterity implies that frontline staff divide their time between the two conflicting activities of exploitation and exploration, and the role of top management is to develop the appropriate organisational context in order that this will happen (Birkinshaw and Gibson, 2004, p. 50).

The underlying problem for which organisational ambidexterity can be forwarded as a solution can in itself be seen as an organisational paradox or a complex set of individual and organisational tensions and conflicting requirements (Andriopoulos and Lewis, 2010). In the midst of the situation, it can for the participants seem impossible to square the circle of both ensuring that there is a relatively stable support structure within the organisation whilst at the same time the organisation seeks to organise itself as a flexible network able to seek and act upon emerging opportunities (Holmberg and Mathiassen, 2001). Establishing contextual ambidexterity can be seen as an example of the transition from a situation of conflicting environmental demands to one of concurrently engaging in exploitation and exploration, and as such represents the fourth type of resolving an organisational paradox suggested by Poole and Van de Ven (1989).

The following section considers in more detail the paradoxical relationship between the user and technological affordances and mechanisms. This is primarily done by highlighting the paradoxical aspects of consumer-product relationships (Mick and Fournier, 1998) and relating such paradoxical aspects to the use of information technology in mobile working. The discussion also draws upon the general paradoxical character of technology in use as characterised by Arnold (2003). The following section discusses the resulting contrary technology performance in terms of competing organisational goals.

3.5   Performance

Every technology is both a burden and a blessing; not either-or, but this-and-that. (Postman, 1992, pp. 4–5)

There is no guarantee that engaging with mobile information technology will lead to the harmonious relationship Weiser (1991) anticipates. In practice, the technology will impose demands, for example, of everyday maintenance tasks, for back-ups to be made, ensuring that batteries are charged and that all necessary cables and connectors are brought along (Sørensen and Gibson, 2008). However, the use of technology can also more comprehensively be seen as conflicting and paradoxical, which will be elaborated on in this section.

Conflicting relationships

Considering the relationship between the technology and the user, there are substantial conflicts at play. For example, the individual user is through their mobile information technology both free to roam and at the same time instantly available for interaction requests (Arnold, 2003). Understanding the processes of adaptation between technological affordances, organisational mechanisms and human activities is a complex arrangement of mediated and body-to-body interaction (Fortunati, 2005). Closing the physical gap between the user and their computational support for interaction increases the complexity of the relationship. Mediated interaction gains immediate access to an individual's personal context of conflicting demands, purposes, preferences, interdependencies between colleagues, performance measurements and the degree of individual discretion in decision-making. A closer physical relationship forged between user and mobile information technology implies a greater need – and possibly also a greater desire – for hospitality and care for the technology (Ciborra, 1997) as affordances and mechanisms are enrolled into and simultaneously shape working practices. Leonardi (2011) characterises this in terms of the imbrication of human and material agencies, and documents how the perception of affordances leads to changes in social practices and perceptions of constraints to technological changes.

Arnold (2003) characterises the use of technology as a non-linear and paradoxical translation from affordances to performances. Technology performances are the results of affordances – possibilities for action – meeting inherently contradictory situations. This contradicts the traditional assumption of a linear and simple relationship where performances apply affordances as simple means of serving specific purposes. Arnold states:

> a certain technology applied in a certain way in a certain context may have consequences or implications of one kind, and may necessarily and at once be implicated in a contrary set of consequences or implications. It is not simply that the performance of the technology falls short of expectations, or that it has unintended consequences, common though both of these are. The claim here is that the difference the technology makes may be a result of a simultaneous move in both directions along any given axis, and the consequences or implications of this movement may not form a single coherent set, but may result in the co-establishment of contrasting conditions. (2003, pp. 231–2)

The categories of affordances and scripts outlined in Chapter 2 denote tensions between contradictory choices where both aspects are potentially valid – for example, stationary versus portable. The portability of a mobile phone affords it to be carried around by its owner, but it also equally affords it accidentally being left behind or indeed purposefully stuck to a dashboard to ensure that it does not get lost. Affordances and constraints characterise the opportunities for potential action or the restriction of such action, and implicitly assume the appropriate context and user. The portability of a PDA assumes both the ability and desire to carry the technology.

Similarly, Mick and Fournier (1998) argue that consumption in general is associated with paradoxes. The closeness of parts of the technological arrangement to the user's body makes it all the more critical as the technology follows the user in their pockets, handbags, etc., and as the user follows the technology. Jarvenpaa and Lang (2005) apply a revised version of the eight paradoxes suggested by Mick and Fournier (1998) in a discussion of user-technology paradoxes relating to how mobile information technology: empowers or enslaves; creates dependence or independence; fulfils or creates needs; offers new competences or incompetences; supports planning or improvisation; engages or disengages; is public or private; and leads to illusion or disillusion. It is argued that users faced with the need to resolve these contradictions will develop coping strategies. See also Table 3.1. This complex and paradoxical relationship between user and technology results in the user inventing and cultivating his or her own individual coping mechanisms in order to resolve everyday conflicts (Jarvenpaa and Lang, 2005).

The same mobile handset that offers an employee easy access to flexibly co-ordinate activities with a colleague on the fly can also be an instrument of instant control and micro-management. Mobile information technology both allows flexible management of personal arrangements and for working life to encroach on the private sphere. The specific outcomes will depend on a combination of organisational practices and regulations, as well as situated improvisation by participants engaged in mobile work. When confronted with the conflicting demand of not disturbing a meeting and yet remaining available for interaction, the mobile phone may be set to silently vibrate, and the user may here choose to interact through text messages instead of calls.

The mobile phone both facilitates the user's mobility and provides a uniquely fixed point of contact: 'We can move, but we are always there' (Arnold, 2003, p. 243). It can render even the busiest person instantly available, yet provide an

Table 3.1 Comparison of Jarvenpaa and Lang's (2005) and Mick and Fournier's (1998) paradoxes. Please note that the last four categories do not match across the two articles

Jarvenpaa and Lang	Mick and Fournier
Competence/Incompetence	Competence/Incompetence
Fulfils Needs/Creates Needs	Fulfils Needs/Creates Needs
Engaging/Disengaging	Engaging/Disengaging
Empowerment/Enslavement	Freedom/Enslavement
Planning/Improvisation	Control/Chaos
Illusion/Disillusion	New/Obsolete
Public/Private	Assimilation/Isolation
Independence/Dependence	Efficiency/Inefficiency

easy means for even the most available person to appear busy. The user is both liberated and tethered by the technology, which at the same time is deeply private and yet often introduces conversations into public events (Arnold, 2003). Technology performances emerge when technological affordance is socially situated, thereby being thrown into a context of contradictions. Performances emerge from situated improvisation informed by individually cultivated coping strategies and can be stipulated by social and organisational practices (Schmidt, 1999).

The primary analytical focus of this book is on the complex interplay between services and performances. Even the simplest of technology affordances can be turned into highly complex and contradictory performances as a result of the richness of individual, social and organisational diversity in which the services are evoked, cultivated and re-shaped (Haddon *et al.*, 2006; Ling, 2008).

3.6   Asymmetry

The tension between technological assumptions of unfolding interactivity and the individual and social expectations or demands of how the interaction should unfold forms a particular strong paradox in the use of interaction technology.

This relates to the debate concerning interaction technology as maps – resources for reflexive actions (Suchman, 1994) – or as scripts stipulating activities (Schmidt, 1999) as outlined in the previous chapter. Here the notions of technology symmetry and asymmetry are introduced. This section presents interaction symmetry and asymmetry as the social practices into which the technological properties will be enacted, rejected, questioned and transformed.

Research on the use of information technology discusses, for example: asymmetry in awareness of a video-mediated presence (Heath and Luff, 1992); asymmetry in interaction preferences (O'Conaill and Frohlich, 1995; Nardi and Whittaker, 2000; Kakihara *et al.*, 2005); asymmetry in the negotiation of desktop video conference interruptions (Tang *et al.*, 1994); asymmetry between the initiator and recipient of synchronous interaction (Wiberg and Whittaker, 2005); and the temporal arrangements of technology and work in organisations (Lee, 1999).

Asymmetry in life

Goffman (1959) suggested that social relationships can be characterised in terms of the asymmetry of everyday interaction – the differences between how we wish others to see us and how in turn we expect to engage in interaction with them. Goffman argued for an asymmetry in this eternal game of impression management where people in general are better at guessing the opinions of others through observation than they are at hiding what they really think of others. Social interaction can therefore be characterised as an endless and asymmetric process of presenting oneself and assessing others' opinions. Face-to-face discussions are based on rich repertoires of subtle cues and mutual adjustments forming interaction ritual chains (Collins, 2004). By projecting the general socio-logical considerations onto the level of individual pragmatic preferences, these rituals also serve the purpose of carefully negotiating and mutually adjusting possible conflicts between peoples' desires and preferences to interact (Ljungberg and Sørensen, 2000; Nardi and Whittaker, 2000).

This asymmetry between our inner disposition and outer manifestation can, for example, explain the social awkwardness and changed practices when peo-ple interact through mobile as opposed to landline phones. Receiving a call on a mobile phone easily leads to a self-inflicted pressure to answer in order to demonstrate a positive attitude towards the caller. Failing to answer may lead to the call recipient feel a need to explain themselves at the next point of interaction between the two. This is a significant departure from situations where the shared assumptions about availability are directly related to specific contexts of interaction defined either by planning or by expectations that people will not be instantly available.

Asymmetry at work

In the context of work, the traditional social conditioning of interaction is extended through mutual interdependencies (Karsten, 2003). Here interper-sonal interaction mostly consists of relatively short, opportunistically organ-ised discussions (Whittaker *et al.*, 1997). Although the literature on meetings tends to characterise these as planned, lengthy events involving a number of people, empirical research has verified that the majority of meetings can be characterised as a series of very brief, unplanned encounters between two indi-viduals discussing ongoing issues and frequently centred around shared docu-ments (Whittaker and Sidner, 1996; Wiberg and Whittaker, 2005).

Goffman's (1959) theatre metaphor characterises how collocated colleagues are socialised into understanding the careful and fragile negotiations of what is acceptable behaviour when only co-workers are present (backstage) and what is acceptable when facing customers, clients or other outsiders (frontstage). General social skills form an essential basis for assessing and acting in these situations. Socialising often takes place in specific physical contexts where all senses can be applied to help assess the intentions of others. In these situations the necessary trust to be able to relax and to let down one's guard backstage is established (Olson and Olson, 2000; Dubé and Robey, 2009). Socialising back-stage can serve as a crucial element in transferring essential knowledge between colleagues, even between mobile workers who would only meet up to share a cup of coffee once each day (Orr, 1996).

When all or a significant part of work is conducted remotely, there may be far fewer opportunities to establish the necessary trust enabling backstage behaviour. This can result in confusion regarding how to interpret the status of meetings, with some perceiving the situation as backstage and others perceiving it as frontstage work where it is important to carefully assess one's actions (Whittle, 2005).

From a pragmatic perspective, mutual interdependency between collaborators does not imply common preferences regarding the practical arrangement of its negotiation and resolution (Ljungberg and Sørensen, 2000). A brief email conversation may easily suffice for one of them, whereas the other may wish for a long meeting. One may wish to discuss matters on a mobile phone immediately, another to postpone the discussion until the following day, while a third person may not see the need for a discussion at all. This is one of the reasons distance matters, as co-presence makes these kinds of negotiation about how interaction should unfold much easier. Here the rich repertoire of human social skills can be mobilised to ensure that any conflict is addressed and resolved in a satisfactory way.

When work situations are distributed and constantly shifting but are instantaneously connectable through mobile services, the natural signalling of individual preferences must be subjected to explicit negotiation through the mobile service. Individuals will seek to either make themselves highly available to ensure relevant interaction is dealt with or will protect themselves from unwanted interaction. Interaction asymmetry here represents the cultivation of interaction priorities. What for one person is a reasonable interruption is considered a rude attempt to disturb by others. Each individual will in his or her own practical context seek to negotiate and communicate both general and specific preferences to ensure that people in their surroundings will align themselves as far as possible to the preferred configurations of subjects to be discussed, the people to engage in discussions with, as well as the time, place and context in which this is done.

The affordance of interaction symmetry presents the paradox of freedom to instantly reach others and at the same time the risk of enslavement to constant connectivity and reachability (Mick and Fournier, 1998; Jarvenpaa and Lang, 2005; Cousins and Varshney, 2009). This implies that coping strategies are employed in order to manage the performances. For example, social and organisational coping strategies involve implicitly conditioning the expectations and behaviour of others through behaviour and discreet hints as well as explicitly through articulating individual preferences and organisational principles.

The elusive fluid working arrangement of a person's surroundings implies the perfect orchestration of people to engage exactly as and when desired. This is to some extent possible for a selected elite of people with substantial organisational status, who can use their influence to orchestrate others around them. However, as most work activities will to some extent be governed by uncertainty and since most people will not have extended powers to orchestrate their interaction contexts, everyday working life can be described as the continuous individual attempts to cultivate particular asymmetrical interaction arrangements suiting their specific needs and desires. These efforts will be implemented in the uncertain context of others also seeking to obtain fluid interaction patterns through their own asymmetrical preferences, and what for one person is a nicely fluid working day can for others be a day of constant interruptions.

3.7   Boundaries and fluidity

Mobile work is conducted under organisational paradoxes, contradictions, tensions and conflicting requirements. Mobile workers consequently engage in complex arrangements of planned interventions and emerging decision-making (Quinn and Rohrbaugh, 1983) and deal with these through balancing planned and emergent technology performances. This involves improvisation (Leonardi and Barley, 2010, p. 28) and shifts of routines and technological configurations (Leonardi, 2011).

This book's main assumption regarding the role of enterprise mobility is that planned and emerging technology performances serve an overarching challenge of continuously resolving the paradox of demand for both fluidity and boundaries. The distributed application of mobile services provides rich opportunities for the fluidisation of, for example, interpersonal interaction, the arrangement of responsibilities in collaboration and the control of highly distributed activities. However, mobile technology performances also support the formation of new individualised boundaries for interaction, boundaries in collaborative arrangements and organisational decision flows. The technology performances signify the contextual appropriation of mobile services drawn from diverse portfolios of services characterised by portability, connectivity, pervasiveness, intimacy, priority and memory. Figure 3.2 illustrates these challenges of cultivating fluidity and boundaries through planned and emergent technology performances – the meeting of mobile work paradoxes and mobile services portfolios.

A simple relationship between technology performances and the cultivation of fluidity and barriers cannot be assumed *a priori*. Some boundaries are the result of planned technology performances. Equally, fluidity will to some extent be the result of emergent performances. However, the flexibility of mobile services in action can also result in boundaries created decentrally at the spur of the moment. Equally, fluidity can be planned and facilitated.

Here we investigate the cultivation of fluidity and boundaries through planned and emergent technology performances in the context of *creativity*, *collaboration* and *control*. *Creativity* is in this context defined as seeking to understand the mobile worker's technology performances serving the purpose of managing a range of conflicting opportunities, desires and demands. *Collaboration* emphasises technology performances aimed at resolving mutual interdependencies

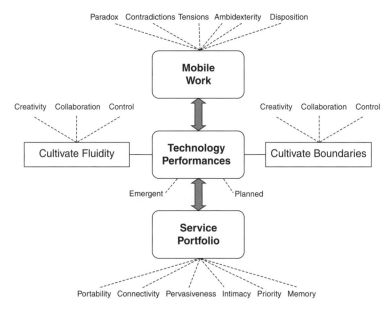

Figure 3.2 The cultivation of fluidity and boundaries as the core paradox of technology performances in enterprise mobility

with others and in meshing mobile collaborative activities. The study of *control* is concerned with technology performances serving the purpose of controlling, managing, planning and overseeing mobile work activities.

While it is tempting to extrapolate these three analytical categories into traditional social categories for organisational action, such as individuals, teams and organisations (Jessup and Robey, 2002; Sidorova *et al.*, 2008), it has been the distinct aim of this book to anchor the analysis in individual technology performances. The main unit of analysis is the individual mobile worker viewed in the context of his or her work. The mobile worker creatively manages information, interaction and availability. He or she engages in collaboration with others to negotiate mutual interdependencies in their respective work. He or she also engages in or is subjected to the control, management, planning and evaluation of work activities.

The following three sections will in turn discuss these three aspects before Chapters 4, 5 and 6 analyse an aspect using a series of nine case studies.

3.8 Creativity

This perspective considers that the individual mobile worker's information management and interaction creativity is required to manage conflicting requirements, disruptions, over-constrained environments and increasing work pressures. Mobile information technology performances afford fluid and creative arrangements of individual interactions according to the need to balance the situation with the individual disposition and mood. However, performances can equally require effort to be invested in managing unanticipated consequences of this interactional fluidity as others seek to resolve their tensions through interaction.

The physical processes of engaging the technology can result in *coupling*, which implies that the activities render intentions effective in a process whose focus is beyond the technology (Dourish, 2001, pp. 138ff). Coupling and decoupling is constituted by a wide range of elements spanning the specific characteristics of the technology, the routines of the user, specific idiosyncratic choices by the user, the actualities of the situation, the norms, values and rules governing the role played by the user, and the institutional context in which the technology is residing. Coupling can be instantiated by others seeking to interact or by the user himself or herself, or can be triggered by stored alerts within the technology. It is unreasonable to assume that the patterns of the user coupling and decoupling the technology are either entirely random or fully governed by some external logic. Rather, they are the result of continuously cultivated routines seeking to address tensions, for example, between the need for emergent action versus planner intervention.

Individual usability

For decades HCI research has studied intimate technology performances primarily with the aim of understanding how to improve the usability of the technical affordances through consideration of the immediate context of the user and the device. Most HCI research emphasises radical prototypes pushing the technological opportunities at the cost of less emphasis on everyday requirements for support (Newman, 1994; Whittaker *et al.*, 2000). However, some work seeks to understand the requirements for everyday use of technology and, for example, suggests that the field should seek to formulate a set of reference tasks (Whittaker *et al.*, 2000) and emphasises the role of context for mobile information technology support (Dix *et al.*, 2000).

However, the discussion in this book is guided by a broader perspective than the traditional HCI discourse, which only rarely explores the organisational context of technology use and how this relates to individual practices (York and Pendharkar, 2004; Sørensen and Al-Taitoon, 2008).

The promise of mobile interaction

The use of mobile information technology implies the promise for the individual of both reaching beyond and of being reached from beyond the immediate context. This amounts to a promise of the compression and fluidisation of space, time and context of interaction (see Figure 3.3) (Kakihara and Sørensen, 2002).

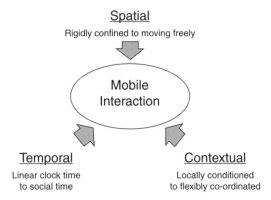

Figure 3.3 The spatial, temporal and contextual opportunities of mobile interaction

Space

Transcending geographical barriers for interaction is associated with the promise of more flexible access to interact with others and information resources.
Interaction is no longer strictly bound to and locally conditioned by the geographical boundaries of the immediate work environment and can be conducted
across spatial barriers. This promise challenges the established understanding of
social interaction (Ling, 2008) as traditionally analysed as co-located (see, for
example, Goffman, 1959; Collins, 2004).

Time

Mobile interaction also challenges the notion of linear clock time and supports social time with more flexible time disciplines (Lee and Liebenau, 2002).
The promise of instant access to colleagues can make for speedy decisions.
However, as argued by Barley (1988, p. 125), the temporal order of the workplace serves not only as a structural template for organising the behaviour of
participants but also as 'an interpretive framework for rendering action in the
setting meaningful'. The structural aspects of temporality are largely measured in clock time by objectified parameters such as sequence, duration, rate
of recurrence and temporal location. The interpretive aspects of temporality
denote how people in the workplace in social time interpret the changes to the
structural parameters.

Based on Hall's (1959 and 1962) work, both Barley (1988) and Lee (1999)
discuss the temporality of work in terms of monochronicity, where people seek
to structure activities by allocating specific timeslots to each event, and polychronicity, where less importance is placed on the divergence of the structural
and interpretive aspects of the temporal order.

Mobile interaction can alter timekeeping by replacing scheduling with direct
micro-co-ordination (Ling, 2004, p. 58). The constant interweaving of situated

and mediated interaction with people, services and technologies can shift the relative importance of time and space in rhythms of interaction. In mobile interaction, *when* a person is available can be more important than *where* they are available (Green, 2002).

Context

Mobile interaction alters the broader contextual aspects of interaction from being locally conditioned to flexibly co-ordinated by the participants, for example, in terms of who is engaged in the interaction, their moods, cultural context and mutual recognition (Kakihara and Sørensen, 2002). Potentially fluctuating individual preferences, organisational practices, power and politics shape the particulars of the relationship fostered between the individual user and the mobile information technology, and result in specific technology performances. It is important to consider the individual user's immediate situation of work beyond a simple 'container' perspective (Sherry and Salvador, 2001), for example, defined by temporal and spatial co-ordinates. Mobile information technology performances reveal these magnificent promises in action and thereby throw light on a more complex reality that is not primarily signified by grand promises but by everyday realities that are just as much shaped by contexts, situations, routines and practices as by the promises.

Container or *Befindlichkeit?*

The interpretation of *context* depends on the explicit or implicit assumptions of the researcher. For the engineer researching 3G base-station optimisation, it may primarily be quantitative measurements reflecting signal strength and reliability that are dependent on the distance between the handset and the base station. For other technical researchers, context may simply equate location according to a technically constructed electronic grid, for example, through GSM triangulation, embedded electronic tags or absolute GPS position. For more sociologically informed studies, context can denote the shared understanding of the situation a person is in, for example, as illustrated by Weilenmann's (2003) recordings of teenage girls exploring the context of friends through discussing what activities they were engaged in rather than exactly where they were.

Here the context of interaction will be defined in line with the arguments of McCollough (2004), who characterises context in terms of 'work practices, social organization, and physical configurations'. It is a way of capturing the totality of the social and physical aspects from an abstract external point of view. Ciborra (2006) applies a phenomenological understanding emphasising the Heideggerian notion of *'Befindlichkeit'* or 'state of mind' to define situations. Ciborra criticises the lack of definition of situation in the literature and argues that the use of the concept generally implies an artificial categorisation of aspects of situations and in particular the separation of physical and emotional aspects. This resonates with Dourish's (2001, p. 107) concept of 'embodied interaction', which is partly derived from Heidegger's rejection of the dualism between body and mind.

Interaction expertise

Fulton Suri (2005) documents everyday examples of the intimacy cultivated using material objects in specific situations and illustrates richly creative everyday innovation. For example, she documents how people: *react* through automatic behaviour with objects in a particular context, such as sticking to the lines at the side of the street when cycling; *respond* in specific ways to objects displaying affordances, for example, using one finger as a temporary bookmark when temporarily engaging in another activity; and *co-opt* by actively taking advantage of the opportunities presented by the environment, such as using the handlebar of a bicycle for the shopping bag. Portable technology lends itself better to the cultivation of intimate everyday use than stationary technology. It follows the individual user throughout his or her everyday activities. The resulting practices routinise the translation from technology affordances into technology practices. Similarly, perceived technological constraints challenge and innovate technological practices (Leonardi, 2011).

Acquiring such creative expertise in mobile interaction can be characterised by Dreyfus *et al.*'s (1986) much-cited model of human learning. It characterises the span from novice to expert, which Flyvbjerg (2001, p. 20) summarises as: *novice*, acting based on context-independent elements and rules; *advanced beginner*, including the interpretation of situational elements; *competent performer*, using goals and plans to guide actions; *proficient performer*, intuitively identifying problems, goals and plans, which are analytically evaluated before actions are taken; and *expert*, engaging in flowing, effortless performance unhindered by analytical analysis. For the expert, 'the separation between person and machine, between subject and object disappears' (Flyvbjerg, 2001, p. 19).

Daily interaction with technology, the close relationship and the importance of technology to accomplish work are likely contributing factors in rapidly fostering expertise in mobile information technology use. Ericsson *et al.* suggest that expertise emerges as a result of intense practice for at least a decade, but their suggestion implies at least 10,000 hours of practice (1993, Figure 12), which translates into an average of less than three hours' practice each day. This is popularised in a number of books, most famously Gladwell (2008). The daily use of mobile information technology during a 40-hour working week implies that any user with little more than five years' experience in using the technology by this measure could be an expert – and only three years' experience in cases of continuous cultivation of mobile phone use throughout the waking hours of the day.

However, Sennett (2008, p. 9) argues that: 'There is nothing inevitable about becoming skilled, just as there is nothing mindlessly mechanical about technique itself.' The 'intimate connection between hand and head' in crafts-manship explored by Sennett is relevant for the understanding of everyday technology performances. Sennett (2008, p. 10) argues that all skills begin in bodily practices, such as those emerging from touch and movement. Continual attention to the management of information and haptic engagement renders the use of information technology a materialised practice. Mobile phone profiles are changed depending on changing situations, emails are constantly checked, address books are updated so that a new caller is automatically identi-fied the next time they seek attention, and a range of other embodied practices are engaged – all by complex gestures.

3.9 Collaboration

Collaboration can be characterised in a multitude of ways – for example, as negotiation and resolution of mutual interdependencies (Schmidt, 1993), information flows (Ciborra, 1993), semiotic translations in the generation and dissipation of information (Ramaprasad and Rai, 1996) or the management of knowledge (Newell *et al.*, 2009).

The analysis of collaboration in this book primarily considers what Goffman (1959, p. 88) described as 'performance teams'. These are flat organisations, an argument also made by Schmidt (1993). Within such performance teams, the analysis in particular considers the individual mobile worker's technology per-formances in order to negotiate and resolve mutual interdependencies and to mesh their relevant work products into a collaborative whole. The chosen per-spective is one of primarily considering the mobile interaction of an individual within the context of collaboration, as opposed to, for example, the structural aspects of teamwork organisation, as in Perlow *et al.* (2004).

Significant research seeks to understand collaboration. For example, McGrath (1991) forwards a general account of the temporal interaction and performance of groups in terms of their functioning, well-being and member support across the activities of problem identification, problem solving, conflict resolution and execution of work. Watson-Manheim and Bélanger (2007) argue that teams co-ordinate, share knowledge, gather information, develop relationships and resolve conflicts. Engeström (2008) explores how teams can collaborate effectively – and become 'knots'. The analytical distinction between horizontal and vertical arrangements of work activities is constantly under negotiation in order for organisations to optimise their behaviour under the given constraints. For example, Perlow *et al.* (2004) explore structuration through patterns of interaction between engineers and project leaders in small groups. Much work has been done within transaction cost theory to understand the dynamics of collaboration (Coase, 1937; Williamson, 1981) and the role of information technology therein (Ciborra, 1993), as well as criticisms of the transaction-cost perspective for understanding work (Schmidt, 1994; Scarbrough, 1995).

Flexible working practices, pressures from competing values, available technological support for remote interaction and other factors imply that *working* together is increasingly disassociated from *being* together. Work activities are always socially and materially situated (Suchman, 1987). However, the demand for rapid and complex decisions has extended the reach and flexibility of these situations through a variety of technology performances. Horizontal and direct collaboration negotiates mutual interdependencies across a multitude of shifting work contexts. This involves the co-ordination, meshing, scheduling and interrelating of individual activities and work products (Schmidt, 1993).

Groupware and CSCW

With the advent of networked personal computers, the notions of groupware (Johansen, 1988; Bullen and Bennett, 1990) and computer-supported cooperative work (Schmidt and Bannon, 1992) emerged as related fields in the mid-1980s. These fields drew together researchers from computer science, HCI and social sciences, primarily the sociology and ethnography of work, as is well illustrated by Baecker (1993). Initially, groupware research tended to assume a high degree of shared goals amongst a small and fixed group of collaborators (see, for example, Ellis *et al.*, 1991). Such assumptions about groups can been questioned (McGrath, 1991; Schmidt and Bannon, 1992). Teams or cooperative ensembles can be: large or embedded within larger teams; transient formations with unstable or non-determinable membership; dynamically changing according to situational constraints; and distributed logically in terms of control and across time and space (Schmidt and Bannon, 1992, p. 15). The understanding of mobile working is an emerging interest within the CSCW and groupware communities, but it has always been a relatively under-researched issue.[10]

Working together

Collaboration serves the purpose of enabling the accomplishment of results beyond the limitations of the individual. Schmidt (1993) characterises this in terms of: (1) *directly augmenting* the physical or cognitive capacity of several individuals; (2) the *differentiation and combination* of technique-based specialisations; (3) *mutual critical assessment* deploying multiple problem-solving strategies and heuristics to solve equivocal problems; and (4) *confrontation* and a *combination* of perspectives, denoting attempts of reach decisions by reconciling multiple incommensurate local and temporary conceptualisations of the situation at hand.

Work varies greatly in terms of interactional complexity and environmental uncertainty. For example, Perrow (1984) evaluates the risks associated with various systems and characterises their complexity in terms of the degree of coupling between the elements (tight or loose coupling) and of the extent to which the system is linear or complex. Woods (1988) provides a similar conceptualisation of system complexity and the following is drawn from his definitions. The work setting may include many geographically distributed actors. It is much more complex to work together with thousands of people scattered across the globe than it is to work only with one person sitting in the same office. The collaborative process will be more complex if the product or service resulting from the work consists of a large number of interdependent activities. Work is also more complex if the collaboration requires significantly different areas of competence with different conceptualisations, identities and priorities. The clear conception of one goal or purpose clearly makes it easier to work together compared to situations where there are strongly diverging goals. Work carried out over a long time-span implies more complex collaboration than work where the outcome is instant decisions without the need for these to be remembered, as the results will be used immediately. At the other extreme, if instant decisions are time- and/or safety-critical, then this can be a source of complexity.

Schmidt (1993) suggests dividing the complexity of collaboration into: (1) structural complexity, denoting the degree of interactional complexity; (2) temporal complexity, relating to the dynamic behaviour of the system and the extent to which tight coupling results in time-critical decisions; and (3) apperceptive complexity, where perceiving and interpreting the state of affairs may be hindered, for example, by noise, unreliable sensors and ambiguous information.

Mutual interdependencies in collaboration

The individual actors engaged in collaboration become mutually interdependent, as through the process of working together they will critically rely on others. The concept of interdependence is therefore at the core of understanding collaboration (Schmidt and Bannon, 1992; Schmidt, 1993; Karsten, 2003).

Thompson (1967) suggests the concept of 'internal interdependence' and describes this in terms of the following categories: *pooled interdependence*, where each sub-part contributes to the whole; *sequential interdependence*, which denotes an assembly where B relies on the actions of A; and *reciprocal interdependence*, where the output of each becomes the input for others. Schmidt (1993) argues that Thompson's conceptualisation of interdependence is problematic as a basis for understanding the operational aspects of collaborative work. The concept is developed with a firm's financial performance as the unit of analysis. Further he argues that all collaborative work is mutually interdependent. If A and B work together and B relies on the output from A as the input of his or her activities, then we cannot assume the strict unidirectional relationship that Thompson indicates. B may rely on A in terms of the quality, timeliness and quantity of delivery, but A will also be critically dependent on B in terms of feedback on the quality of the work delivered. Thompson fails to include other types of interdependencies than those that are measurable.

It is essential to distinguish mutual interdependence in collaboration from the issue of multiple persons sharing a common resource. It is, for example, not mutual interdependence when a number of drivers negotiate the same stretch of motorway or when cheap airline tickets are sold to a group of customers. From each participant's point of view, he or she would probably prefer to have the entire road or airplane to himself or herself. Others are in this respect an added complication that can be done without. However, the inclusion of others in collaborative activities is, as we saw above, desired precisely because it makes it possible to solve problems beyond the reach of each participating individual.

As argued by Karsten, the issue of interdependence can be further expanded to take on a variety of meanings. She discusses interdependency creation as social practices and argues:

> The relationships are not considered as separate entities, but the attention is on people engaging in action and interaction. Interdependencies are then seen as constantly constructed and reconstructed social practices ..., that is, repetitive, patterned, and reciprocal action and interaction ... Interdependence construction is then the creating or reconstructing patterns of action and interaction where two or more people are mutually dependent on each other. (2003, p. 2)

Karsten (2003) applies Giddens' structuration theory and suggests four aspects of interdependence construction: *social integration*, which is the systemness of densely reciprocal interaction between actors; *system integration*, which signifies the systemness of interaction between social systems; *time-space distanciation*, which is the persistence of a social system; and *institutionalisation*, where relationships become routine through discernible patterns of interaction.

The broad range of concerns from Thompson's (1967) discrete view of interdependence in terms of financial performance, through Schmidt's (1993) perspective highlighting the operational aspects of collaborative activities, to Karsten's (2003) broad understanding of interdependency construction as social practices demonstrates the definitional variety of interdependencies. The following sections and Chapter 5 will adopt the perspective suggested by Schmidt, emphasising an operational understanding of the functional aspects of interdependencies.

Conceptualising collaboration

Cooperative work arrangements consist of mutually interdependent actors engaged in collaboration through their particular division of labour. The cooperative work arrangement is used similarly to the concept of application domain within systems analysis and design (Mathiassen *et al.*, 2000). Collaborators interact with, control and transform their common field of work, which consists of mutually interacting objects and processes (Schmidt, 1993; Schmidt and Simone, 1996). The common field of work defines the locus of the mutual interdependencies – the commonality of the collaboration. It is a conceptual structure supporting the analysis of the collaboration and is not a tangible and directly observable phenomenon, although essential elements in the field of work may be tangible. The common field of work is similar to the notion of problem domain (Mathiassen *et al.*, 2000) in systems analysis and design.

Collaborative work activities are characterised by the interdependent actors engaging in changing the state of objects and processes in the common field of work (see Figure 3.4). The field of work is the exact common focal point of the interaction and the cornerstone of the mutual interdependencies in the collaboration. Two people sitting in an office next to each other may or may not be part of a common field of work. This conceptualisation of cooperative work assumes an operational perspective as opposed to studies emphasising economic ownership or organisational affiliation (Schmidt, 1994). This makes

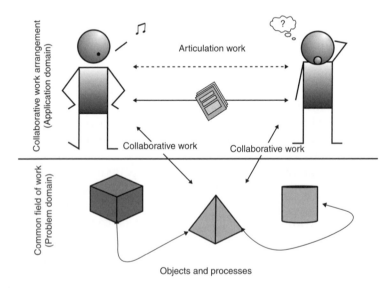

Figure 3.4 The articulation of mutual interdependencies in collaboration: illustrating the main elements in collaborative work, drawing concepts from both Schmidt (1993) and Mathiassen *et al.* (2000)

it particularly suitable for the analysis of highly distributed and mobile working, where the flow of resolving mutual interdependencies crosses temporal, spatial and contextual barriers.

Articulation work

In order for the actors to engage in collaborative work by altering the state of the field of work, they will need to co-ordinate their activities and to understand the current state of affairs and the roles of the various actors in this (Schmidt, 1993). This can only to a limited extent be done by actors directly observing the actions of others and their results. When the complexity of the work increases ever so slightly beyond the most trivial setting, it is no longer sufficient to observe in order to ascertain the current state of affairs. Schmidt (1993) illustrates this with the case of the hot rolling mill, where four people work around the same machine in order to roll glowing metal plates into rods. Even with only four people, the noise made by the machine increases the (apperceptive) complexity, so they need to shout and otherwise indicate to each other in order to co-ordinate their movement of the hot metal whilst operating the rolling mill. Fields of work more complex than this require explicit co-ordination of the activities performed. This kind of work, namely symbolic utterances about the work, has been characterised as *articulation work* (Strauss, 1985; Gerson and Star, 1986).

Strauss (1985) argues that the articulation of work is a type of work in itself and Schmidt (1993) suggests that this articulation can be in terms of the contextual aspects of: actors; responsibilities; tasks planned; activities performed; common resources; and conceptual structures used within the community. Articulation work is always conducted in relation to the state of the field of work, the demands of the work environment and the wider organisational setting, as well as to temporal and spatial references.

Articulation work can be conducted by directly referring to the field of work – for example, by an actor pointing out or positioning objects as a sign for others to notice. However, articulation work often represents symbolic references to the field of work. Such acts do not alter the state of affairs directly. Articulation work can be conducted in more or less obtrusive manners, i.e., the extent to which the act requires the direct attention of others. Finally, articulation work can be ephemeral, in the sense that no inspectable or persistent traces are left behind.

An essential aspect of co-located working involves collaborators establishing mutual awareness of the activities of others (Heath and Luff, 2000). The systemic complexity of articulating the cooperative activities in mobile working is generally higher than collaboration in local or remote working (see Figure 3.1), as temporal, spatial and contextual uncertainties make it more difficult to obtain an overview of the situation.

Mobile workers engage in meshing work by resolving mutual interdependencies in collaboration. They engage in both planned and stipulated performances. Planned performances stipulate behaviour expressed in a variety of statutes, rules, regulations, forms and paper- or computer-based co-ordination mechanisms (Schmidt, 1993; Schmidt and Simone, 1996).

3.10   Control

Control is here primarily understood as the activities involved in overseeing and managing work by balancing planned intervention and emergent action through a combination of stipulated and emergent technology performances. This involves balancing the individual mobile worker's discretionary decisions to engage emergent and planned technology performances. It implies the balancing of contradictory forces of either providing excessive discretion that can hamper collaboration or engaging in counter-productive micro-co-ordination, both of which are so easily afforded by mobile information technology. Control also engages activities aimed at managing internal and external organisational boundaries of interaction.

Controlling distributed work

The issue of organisational control is a core aspect of management and is a wide-ranging issue that has been subjected to decades of extensive research. Control can be defined as: 'Influence of one agent over another, meaning that the former causes changes in the behaviour of the latter; and purpose, in the sense that influence is directed toward some prior goal of the controlling agent' (Beniger, 1986, p. 7). Control can take on many forms, such as direct control studied by Kunda (1992) in project organisation or indirect normative control (Willmott, 1993; Robertson and Swan, 2003).

The ability to control is closely associated with the use of a variety of information technologies, as discussed in great detail by Beniger (1986) and Yates (1989). At the end of the nineteenth century scientific management resulted in extensive innovation and use of communication technologies facilitating remote control through standards and mechanisms (Yates, 1989). Much collaboration continues to rely intimately on direct observation and mutual adjustment (Mintzberg, 1983). However, with the distribution and individualisation of work and the use of mobile information technology, direct observation and mutual adjustment must be conducted through information services.

The complexity of contemporary corporate socio-technical systems challenges the traditional understanding of management as control (Ciborra and Associates, 2000, p. 4). The meeting of top-down strategic decisions and bottom-up adjustments to emerging situations creates drift, where changes and decisions cannot directly be attributed to top-down decisions.

The use of mobile information technology relates in a complex and contradictory manner to goal incongruity and performance ambiguity (Wiredu and Sørensen, forthcoming). Greater connectivity can enforce the stipulation and control of distributed activities. Yet extending organisational infrastructures to mobile workers at the edges of the organisation can also support local improvisation to meet the emergent need for decisions, possibly resulting in less control.

Ciborra provides an example of a source of drift:

> Drifting can be looked at as the outcome of two intertwined processes. One is given by the openness of the technology, its plasticity in response to the re-inventions carried out by users and specialists, who gradually learn to discover and exploit features, affordances, and potentials of systems. On the other hand, there is the sheer unfolding of the actors' being in the work flow and the continuous stream of interventions, tinkering, and improvisations that color perceptions of the entire system life cycle. (2002, p. 87)

Control relates immediately to planned technology performances in terms of the ability of the controlling agent to control behaviour through enforcing stipulated planned performances evoking specific technology performances. However, control also relates to emerging technology performances, for example, in terms of interactivity allowing micro-cycles of remote observation and the resulting controlling actions. Drift results from the need to reconcile conflicting requirements through planned interventions and emerging actions.

Travelling salespeople have traditionally been managed by outcome and have enjoyed relative freedom when on the road, as they were difficult to reach. Many office workers managed by process have, on the other hand, been subjected to more direct control whilst in the office. Mobile information technology can now render travelling salespeople available, while flexible working may lead to less direct control of office workers working from home. Changes in working arrangements and technology support also challenge the feasibility of managing the process as opposed to the outcome (Sproull and Kiesler, 1993, pp. 103ff). Little work has been done thus far to understand the management of homeworking (Felstead *et al.*, 2005, p. 120). Even less research seeks to understand the management of mobile working.

Management as recursion

The horizontal negotiation of mutual interdependencies can be analytically distinguished from the vertical monitoring, control and general management of these horizontal activities. Figure 3.5 illustrates how the co-ordination of mutual interdependencies is a recursive phenomenon, in that seeking to

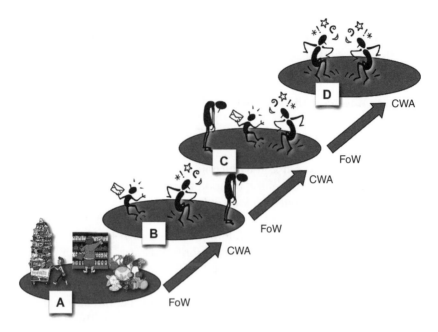

Figure 3.5 The recursive characteristics of the common field of work (FoW) and the collaborative work arrangement (CWA)

manage collaboration in itself can be a collaborative activity (Schmidt and Simone, 1996, p. 159). Consider the situation where the field of work (A) is common to participants in collaborative work arrangement (B). This combination of (A) and (B) will form the common field of work for collaborative work arrangement (C) for the purpose of management, planning and control activities. In (C) separate mutual interdependencies will need to be negotiated and resolved. This implies a vertical nesting of concerns where the cooperative work arrangement and common field at one level both become the common field of work at the following level.

Consider the collaborative work arrangement (B) running a restaurant (A). The members of the arrangement are bound together by their common field of work to order, cook and serve food, pour wine, do the dishes and clean the floor. They will also articulate their interdependencies through talking, shouting and nodding to each other. They will pass orders for food and wine and they will negotiate the relative timing of all these activities. For the managers of the restaurant (C), all of this (A and B) is the common field of work binding them together through mutual interdependencies concerned with deciding the rota stipulating who works when. They will discuss promotions and the general human resources issues, and they may decide the menu with the chef. For the board members of the company (D) owning and running a chain of restaurants, the common field of work is primarily comprised of the management team at each of the restaurants (C), but in principle also includes the floor staff (B) as well as the meals, tables and bottles of wine (A). What for one organisational role is part of the cooperative work arrangement can comprise another role's common field of work. These levels of recursions define levels of management and control. A person in a specific cooperative work arrangement can turn into part of the common field of work to be managed.

Organising teams

The discussion of management as recursion work merely explains that management work implies managing people (also referred to as 'human resources') engaged in collaborative work and can be used to functionally characterise the vertical separation of concerns. One of the essential differences between a hierarchical and a flat mode of organising work is the distribution of roles and the access between the roles. A traditional functional hierarchy is one applying strict principles at each level, seeking to hide the internal operations of the cooperative activities and their articulation two levels down in the hierarchy. The strict division of labour can ensure that the executives at the top will delegate all decisions needed down the chain of command and will only be informed about the operations at an aggregate level. Although this way of organising distributed activities can work very well for relatively stable and simple activities, it has shortcomings when the collaborative complexity increases and work becomes more complex, or if decisions are made in conditions of uncertainty (Mintzberg, 1983).

It can be argued that there are two types of control in collaboration: vertical control, as discussed above, emerging from the recursive character of cooperative activities, and horizontal control, emerging as a means of control in collaboration – this is in essence part of the articulation of work. According to Batt and Doellgast (2005, p. 152): 'Teams set their own sanctions and rewards (horizontal control), while management information systems identified quality errors, flagging good and bad performers with "traffic lights" (vertical control).' This relates to Mintzberg's (1983, pp. 99ff) distinction between horizontal and vertical decentralisation. Horizontal decentralisation implies dissolving some of the managerial responsibilities from the role of being manager, while vertical decentralisation implies dispersing formal power further down the hierarchy of roles in the organisation.

Perlow *et al.* (2004) discuss the design of decision-making in project groups and suggest three patterns of arranging collaborative activities: the hierarchical managerially-centred, the expertise-centred and the team-centred. Each of these signifies designs of group interaction within an organisational and

institutional context. Each design carries with it consequences for the relative distribution of time between engaging in cooperative work activities and
engaging in the articulation of work. The expertise-centred and team-centred
approaches were found to result in a doubling of the time spent on articulating work activities compared with the managerial-centred approach (Perlow
et al., 2004).

Formal versus informal collaboration

Several authors have argued that the increasing uncertainty and complexity of
information work and the horizontal organisation of work activities leads to
increased collaboration and non-hierarchical communication across traditional
organisational boundaries (Hinds and Kiesler, 1995). Informal networks of interaction have always co-existed along with the formal division of labour, even if
these have been viewed as unproductive and have been minimised (Conradson,
1988). However, increased uncertainty and complexity require emergent forms
of organising and complex articulation of interdependencies, bringing with
them the need for informal means of interaction to the foreground (Hinds and
Kiesler, 1995; Kraut and Streeter, 1995). For itinerant experts mostly operating in shifting collaborative arrangements on a contractual basis, networking
activities seeking to establish future engagements become important (Nardi
et al., 2002; Barley and Kunda, 2004).

It is essential to distinguish between the organisation of work in networks
and the distribution of political power and control over the work. Courpasson
(2000) characterises soft bureaucracies as organisations with decentralised
accomplishment of activities and centralised political power and control.
These modern organisational practices replace traditional hierarchies where the
functional relationships across levels in the organisation signify their relative
power, with a separation between the centralised political power and a highly
decentralised conduct of activities. Soft bureaucracies are based on the legiti-
macy of the centralised governance of decentralised activities.

Articulating articulation work

Articulation work serving the purpose of negotiating mutual interdependencies can in itself be subjected to recursion related to control and management
of the articulation of mutual interdependencies – for example, the impromptu
discussion of the meeting agenda in the middle of a meeting aimed at making
decisions regarding sales figures. Here the articulation work related to the common field of work suddenly turns into the articulation of what work should be
articulated: planning. One further recursion would be the principled debate of
how meeting agendas are set and changed. Yet another round could raise the
general issue of decision-making and governance within the organisation.

Making decisions together suddenly is no longer a matter of collaborating; it is the subject matter being articulated. It is no longer an issue of making crucial decisions; it is one of discussing how these decisions are made. This can in turn be a subject for discussion and can result in deliberations of how to arrange a more principled discussion of how decisions generally ought to be made. In this sense articulation work becomes a recursive phenomenon that itself can be made subject to the articulation of cooperative effort (Schmidt and Simone, 1996).

Degrees of local control

The discussion of articulating articulation work relates to the individuals' degree of local control as explained by Schmidt (1993, p. 91): '... there is a clear ladder from, at the one end of the spectrum, modes of interaction that do not involve any pre-specified stipulations – to modes of interaction that involve the prescription incorporated in and the active mediation of pre-specified artifacts, at the other end of the spectrum.' This spectrum, for example, covers: *ad hoc interaction* with no imposed control other than general social conventions; interaction guided by means of specific *conventions*; interaction guided by *written statutes*; the use of *passive mediating mechanisms* such as paper-based forms and organisational charts; and the use of *interactive, computer-based co-ordination mechanisms*.

For most work there will generally be assumed to be a need for a variety of technology spanning the range of degrees of local control from highly restricted activities stipulated by pre-determined or adaptive scripts to ad hoc interaction (Schmidt, 1993; Schmidt and Simone, 1996). Emergent performances engage in ad hoc co-ordination by reconfiguring, overriding or otherwise using a mechanism as a resource of improvised action (Schmidt, 1993). Emergent performances can also evoke open-ended affordances establishing direct connectivity. The distinction between emergent and planned technology performances does not imply any division or roles between those providing the mechanisms for such planned performances and those assumed to follow it. Planned performances can also be pre-organised by the mobile worker who then later either follows the plan or circumvents it.

3.11   Summary

The discussion of enterprise mobility is grounded in issues concerned with flexible, distributed and mobile working. Here the paradoxes and tensions of the conflicting requirements of everyday work demand the balancing of planned interventions and emerging actions. For the individual mobile worker this translates into the meeting of mobile service portfolios and working practices into planned and emerging technology performances. Technology performances can be categorised into: creative management of interaction and information under conflicting demands; collaborative activities in order to negotiate mutual interdependencies; and control activities to oversee, plan and manage mobile work.

4
Creativity – Fluid Performances

Physical proximity to the user is a defining characteristic of mobile information technology. Compared with other organisational information technologies tied to specific workstations, mobile information technology transcends these and follows the worker across his or her workscapes, i.e., the total network of work-places and workstations where the worker conducts their activities (Felstead and Jewson, 2000, pp. 16ff).

This and the following two chapters explore a range of technology perform-ances across different domains of mobile work applying a variety of mobile services. Each of the three chapters presents a number of case studies. Each case is organised as an initial case vignette introducing typical technology performances and mobile working for a particular person, followed by a brief presentation of the context of the case. Subsequently the case is analysed and pertinent theoretical issues are drawn out for further discussion.

This chapter analyses and discusses the creative management of interaction and information in three examples of enterprise mobility: a highly mobile Tokyo CEO; a London Black Cab driver; and a Middle-Eastern mobile foreign exchange trader engaging in mobile trading.

4.1   Case 1: Hiro – CEO

Vignette 1

Hiro is the CEO of a small Tokyo-based company developing digital services for Internet-enabled mobile phones. He walks down the main street in Tokyo's Akihabara district engaged in one of his favourite pastimes: to find inspiration for new services by immersing himself in Tokyo's streetlife. He observes what people do, what they buy and what they wear. Being in the field of the Japanese consumer is a source of inspiration to Hiro, and he characterises this behaviour as 'being analogue', as opposed to surfing the Internet for inspiration. Hiro's com-pany only employs around 20 people and he is the hub of most activities. Whilst traversing through Tokyo, Hiro is therefore subjected to an overwhelming amount of requests for his attention through emails and calls to his mobile phone. His mobile phone is specially customised for him by one of his client companies – a mobile operator that also manufactures handsets. It represents significant financial and symbolic value, as it is the only one of its kind. The 'stealth black' phone is set up with a complex arrangement of bespoke alerts and ringtones depending on who is seeking his attention (study 1 in Table 1.1).

Context 1

Hiro's background is atypical for a Japanese professional as he has never been in traditional employment within a large organisation. He began his career as a successful musician, then he managed an internet service provider (ISP) company in the 1990s before setting up his own company developing mobile services and content in 1998. Hiro sees himself as 'a man of ideas' and a 'producer' rather than a traditional executive (Kakihara, 2003, p. 177). He wishes to create innovations that have broad appeal to people. He seeks to maintain managerial control over his organisation by ensuring it remains small. This provides him with the ability to engage more directly. The organisation is highly successful in developing applications and content for NTT DoCoMo i-Mode mobile phones. Hiro differs from the traditional image of a Japanese executive in the sense that he chooses to work for his own company.

   Peters and Waterman (2004, first edn 1982) popularised the organisational best practice from Hewlett Packard of 'management by walking around'. Senior managers would both exercise their influence and at the same time better understand the current situation within the organisation. By being situated beyond the inner sanctum of the executive offices and by 'patrolling' the outer perimeters of the organisation, executives would remain in touch with and would have influence over the rest of the organisation, for example, through their charisma (Waldman and Yammarino, 1999). Edström and Galbraith (1977) had already argued that the transfer of top managers between branches could facilitate the creation of informal networks supporting the control and co-ordination of work. Hiro's company critically depends on a constant flow of new ideas for mobile applications, of which only a very small number will succeed. Such a high-velocity environment requires a rapid strategic decision-making process, which in turn relies on real-time information, the simultaneous consideration of alternatives and the integration of decisions made by experienced executives. In the fast-moving mobile services sector, Hiro relies on almost continuous interaction in order to gather information and to remotely influence detailed decisions (Eisenhardt, 1989). For him, this is done by a combination of freely roaming Tokyo, studying people to get inspiration and attending critical meetings, while continuously remaining in touch with the rest of the organisation through his mobile phone.

Hiro is highly sought after both by people within the organisation and by a range of external stakeholders. On a typical day he will participate in two to four business meetings and throughout the day he will be contacted on his mobile phone by a number of people. He typically receives 50 emails each day addressed to him directly, i.e., excluding spam emails and messages automatically generated from lists he subscribes to. He is an executive and, as such, the most important part of his work is interacting with clients and colleagues (Hambrick *et al.*, 2005). He defines the company to the outside world and he is therefore faced on a daily basis with the dilemma of either accepting a large number of requests or potentially damaging business relations. He copes with this dilemma by ensuring that as much interaction as possible is managed from his mobile phone. All incoming emails are forwarded to the mobile phone. Due to the large number of messages, the phone is configured so that only messages from contacts deemed important result in a sound notifying Hiro of their arrival. This list of important contacts changes continually and, for example, contains his secretary, members of active projects and a few others. All other emails result in the mobile handset vibrating.

4.2   Intimate technology performances

Hiro's hectic everyday existence requires constant balancing of emerging actions and planned intervention:

> All members in my company and most clients and business partners know that I'm quite busy [laughs]. And they all know that I use mobile phone and email like this [forwarding emails to the mobile phone handset and replying to them as quickly as possible]. So they rarely ring my mobile, except in the cases of really urgent issues. Instead, they send me an email to check if I'm available or not. (Kakihara, 2003, p. 182)

The drift from synchronous to asynchronous interaction has the consequence of storing interaction for later. Synchronous interaction uses time as a filter. Calls missed do not represent actions directly but only indirectly in terms of the social expectations for the caller to be contacted. Using email as instant messaging can be advantageous for individuals in high demand, such as Hiro, as it allows a higher degree of interaction stacking whilst maintaining a rapid turnaround.

Hiro applies his creativity in continuously shaping the filtering properties of the phone to suit his changing needs – both in order to be shielded from requests deemed of less relevance and simultaneously to be made aware of those deemed essential. He shapes the mobile phone as an intimate object of his own by continuous reconfiguration. Through this adaptation it becomes a hub for his interaction. Complex technology performances and coping mechanisms resolve emerging conflicts and contradictions, and exemplify contextual ambidexterity in action. Hiro is simultaneously away and available, and is through this both empowered and enslaved by the technology. The mobile phone makes it possible for him to manage interaction remotely whilst wandering the streets of Tokyo, yet this may in itself increase the volume of interaction requests as he can always be reached. Pre-planned meetings and the technological support for filtering incoming interaction are balanced against the unpredictability of what requests and situations will appear next.

Although mobile email affords asynchronous interaction, in the hands of Hiro it almost becomes synchronous. He argues:

> Many people say that email is asynchronous communication. But I think it should depend on the content of the email. In the case that the content of an email is urgent or important, I reply to it as synchronously as I can. If it's not, I reply to it asynchronously by using available time [...] Currently, I make the decision based on who sends the email. When an email is sent by a really important person, I reply to it immediately, as real-time as I can. But overall, it's important not to leave email unanswered for a long time. Basically I try to answer the emails I receive as quickly as possible. For this purpose, forwarding emails to my mobile is indeed very useful. (Kakihara, 2003, p. 181)

Grudin (2002) emphasises the importance of asynchronous services in ubiquitous computing as essential support for the distribution of information in mobile working. By channelling synchronous interaction to asynchronous messages, Hiro creates the opportunity to apply discretion concerning exactly when he chooses to deal with it. Asynchronicity creates a buffer of peripheral awareness (Schmidt, 2002) of who is seeking his attention and thereby allows him to prioritise.

This convergence of synchronous and asynchronous interaction into one stream of emails or text messages is also illustrated in Mazmanian *et al.*'s (2006) study of BlackBerry mobile email use amongst investment staff in a small private equity fund. In both the cases of Hiro and the BlackBerry-using investment bankers, highly discretionary work across multiple workscapes relies on synchronous interaction being channelled into asynchronous mobile emails. The resulting technology performances are in both cases constant connectivity and near-synchronous responses to interaction requests. Mobile email is characterised by a partner in the investment bank as 'a communication tool of total flexibility' (Mazmanian *et al.*, 2006). The investment managers felt that the intimate technology performances offered the opportunity to monitor information flows and to control the form of information delivery – also key concerns for Hiro's intensive use of mobile email.

Investment bankers display a widespread compulsion to check email and a lack of ability to disengage (Mazmanian *et al.*, 2006). The sense of being in control as synchronous interaction requests are channelled into asynchronous emails results in obsessive behaviour fuelled by creeping expectations of instant response to emails. Here the power of the collective (Mayer-Schonberger, 2009, p. 133) can force the behaviour of individuals by creating drifting obligations towards expectations of instant response. As an example, email can be perceived as a means of engaging with an asynchronous and electronic inbox and outbox. However, in the use of email where participants reply to received email instantly, for example, as they may have instant access through a mobile handset, the technology can be construed by participants as a synchronous means of interaction, turning email into 'realmail'. Over time this transformation has for Hiro also resulted in the asynchronous email medium in effect being used as a synchronous medium without the advantage of time passing as a filter deleting requests not dealt with when they occur. Generally the distinction between synchronous and asynchronous interaction is not determined by technological affordances but is shaped by the norms governing the interaction (Rettie, 2009). As argued by Markus (1994), the social context in which the interaction is interpreted plays a crucial role in its effectiveness. As the CEO and founder, Hiro has significant influence over this context of interpretation.

Both Hiro and the investment managers use email as instant messaging. However, an email client is different from an instant messaging application where the notion of a session precedes the rapid informal interactivity. The establishment of a session applies a priority service and is based on notions of availability and shared context. Such additional 'outeraction' activities are engaged in order to manage the communication (Nardi and Whittaker, 2000). The outeraction of establishing instant messaging sessions represents negotiation, but the email session is always open.

The technology performances documented in these two cases display an intimacy and relentlessness in the relationship forged between the user and his or her mobile email. The relationship is beyond the image of stationary email as a replacement for the letter pigeonhole, where being away would result in messages piling up, such as in the evocative imagery of Palme's (1984) paper title: 'You Have 134 Unread Mail! Do You Want to Read Them Now?' For the contemporary user of mobile email, the challenge is about being on top of the inbox. This denotes a departure from the traditional understanding of email as uniquely asynchronous, where going on holiday meant returning to a packed inbox. However, mobile email follows the user wherever he or she goes. As Carr (2009) expresses the problem: 'Asynchrony, once our friend, is now our enemy. The transaction costs of interpersonal communication have fallen below zero: It costs more to leave the stream than to stay in it.'

4.3   Case 2: Ray – taxicab driver

Vignette 2

It is early Tuesday morning and Ray has just begun working. He owns one of the 20,000 licensed London Black Cabs. As an independent taxi driver, he has extensive discretion regarding where and when he works. Ray has decided to start early today. He is picking up his teenage daughter from school in the afternoon and is taking her to an appointment with their doctor. As he drives down Oxford Street towards Marble Arch, one of his three mobile phones starts ringing. This particular phone is exclusively used for a service automatically locating an available cab nearest to the location of the caller's mobile phone. Ray answers and within five minutes the passenger is picked up at Notting Hill Gate. During the drive, the passenger is busy checking emails on her BlackBerry and also makes a brief call, while Ray is listening to the radio. As Ray drops the passenger off in front of the Houses of Parliament, one of his colleagues calls and informs him that due to a problem with one of the local train lines there is a need for a number of cabs to replace the train for a few hours. As it is good money, Ray decides to accept and sets off to the station. Once done, he decides to have a small break and he finishes by accepting a job posted on his computer-cab terminal (study 2 in Table 1.1).

Context 2

For the past 420 years it has been possible to hail a cab on the streets of London, originally by horse carriage. Since 1851 London Black Cab drivers have been certified according to a strict set of exams, 'the Knowledge', ensuring the driver knows: 300 routes in central London covering 25,000 streets within a six-mile radius of Charing Cross, major points of interest, as well as main roads leading in and out of the capital. When passing the exam, the driver is no longer a 'Knowledge boy' or 'Knowledge girl' and swaps the scooter for the green badge carrying his or her driver identification number and licence to earn a living from driving a Black Cab. Most drivers spend three to four years acquiring 'the Knowledge'. Drivers tend to own their own cab and work has always been conducted in a highly independent manner. The first few years after having qualified are spent learning how to be a profitable driver.

Very little central organisation has traditionally been applied to London cabbies. However, some associations of drivers have used radio circuits with central dispatching. In the 1980s these were replaced with computer cab systems allowing associations of cabs to act in a co-ordinated manner – for example, when bidding for lucrative contracts with large firms. The most common computer-cab systems bring the customer in touch with a dispatch centre that solicits jobs. One of the computer-cab systems automatically links the nearest cab to the calling customer's mobile phone location. Generally the individual cab drivers decide what jobs to accept and will be contracted to accept a certain number of computer-cab jobs each month. Computer-cab technology is still not as widely adopted by London Black Cab drivers as in some other European cities.

Around 500 taxi ranks, the remaining 13 of the original 61 cabmen's shelters and other fixed points around London (public toilets and cafes) have traditionally been the only means for cab drivers to exchange information and socialise. This resembles Orr's (1996) account of the shared coffee room serving as a place for photocopy repair engineers to share knowledge through 'war stories'. Beyond their meeting places, Black Cab drivers have traditionally not had decentralised access to information about job opportunities, the traffic situation, etc. It is therefore not surprising that the mobile phone has emerged as a technology of choice. The mobile phone serves as a tool for getting in touch with the rest of the world whilst driving. Cabbies can receive updates from colleagues on particularly profitable work, essential news about traffic situations or simply social interaction with other drivers. They also stay in touch with their family and friends through the phone. Many cab drivers are motivated to do this line of work by a desire for flexible working arrangements in order to better balance work and family life (Elaluf-Calderwood and Sørensen, 2008).

The transient nature of taxi work contributes to making it an interesting subject of study and a small body of research explores taxi work – for example, on how taxi drivers in various countries manage knowledge (Skok and Kobayashi, 2005); on how taxi drivers assess risk (Gambetta and Hamill, 2005); and the use of GPS-based dispatch systems (Liao, 2003). Across the world taxi work has been one of the first civilian frontiers for emerging mobile technologies, and nowhere is the discussion of technology support versus human skills more interesting than in London, where modern digital technologies have to show their worth alongside a driver who has gone through extensive formal training internalising minute details about the topological aspects of the workscape. In the annual competition between a London Black Cab armed with 'the Knowledge' and a minicab with GPS technology, the Black Cab has so far always come out as the winner, although the gap is getting smaller every year (Elaluf-Calderwood and Sørensen, 2008).

4.4   Mobile and anchored

Observing the drivers in the cockpit of their Black Cab reveals a highly integrated relationship between the car, the driver, mobile information technology and a variety of other supporting technologies. The relationship is cultivated through extensive practice allowing each supporting service to find its rightful role in the daily routines. Mobile phones, the hands-free kit allowing the safe use of phones, the computer-cab terminal, the notepad and pen, the radio and the coffee mug all have their place in the car cockpit and are engaged elegantly when needed or when they demand attention. A Black Cab driver is both mobile and anchored in his or her work. The driver is anchored within the car cockpit, which becomes a permanent workstation carefully organised to fit individual needs and preferences. The driver also makes a living through physical movement by positioning the cab in places with the greatest likelihood of customers requiring the most attractive fare and with the least time spent queuing for a customer. A number of other individual requirements may play a role, such as constraints imposed by family commitments. In Felstead *et al.*'s (2005) physically bound definition, the whole of inner London forms the Black Cab driver's workscape.

Laurier (2004) explores how the car for travelling workers can be constructed as a place of work. Whereas these workers travel as part of their jobs, the London Black Cab drivers' job *is* to drive. Easy access to hands-free mobile phone operation is essential and the confined space is laid out to accommodate daily rituals. Whilst driving takes up most of the attention and thereby reduces the ability to safely manage information, some degree of multi-tasking will often take place, for example, a mobile worker flicking through printed emails while driving on the motorway (Laurier, 2004) or taxi drivers taking notes from a customer on their mobile phone while navigating the cab. The car cockpit is adapted as a place of work where conversations are carried out, paperwork is conducted and electrical chargers feed a multitude of devices, such as GPS navigators, mobile phones and laptops. Cousins (2004, p. 89) reports on the case of a mobile worker insisting on only renting large trucks with four power outlets in the centre console.

Driving a taxi is highly individualised work with few formal interdependencies, and traditional London Black Cab work was conducted as rhythms of looking for work alone, waiting and resting together, and occasionally socialising with customers whilst working. The solitary work of seeking customers is be broken up by fares with the promise of conversation, by periods of waiting in a queue at a taxi rank or by breaks in a cabman's shelter or a cafe. Mobile services altered these rhythms of work. The mobile phone offers the promise of conversations and connected presence (Licoppe, 2004) while being anchored in the cockpit of the cab. The mobile phone and the computer cab system both offer the promise of direct access to remote jobs previously out of reach.

The CEO (Hiro) and the cabbie (Ray) both illustrate the paradox of mobile technology concurrently enabling flexibility and stability. Both are free to roam the streets of Tokyo and London, respectively. However, they are also both constantly available through their mobile phones or at the end of a computer-cab connection. Ray is in motion most of the day but is situated in a stable working environment in which he can organise himself. For Hiro, the sense of stability is embedded within his phone. The contradiction of being simultaneously mobile and anchored can be the subject of a variety of coping mechanisms and resolutions. Felstead and Jewson (2000), for example, argue that the increased use of flexible working arrangements in organisations, such as hot-desking, open-plan offices and shared offices, has paradoxically led to an increase in home offices. Workers who previously relied on their office as a stable anchor in their everyday work compensate for its demise by creating a stable safe haven at home.

When drivers concurrently solicit jobs from customers hailing the cab in the street as well as the cab-computer, possible conflicts can occur, which require a coping strategy. One such example is the driver accepting a computer-cab job and then additionally accepting a hailing street customer if the customer is heading in the direction of the computer-cab job or in the same direction as the cab is taking a parcel.

When Ray engages with the computer-cab system, a computer-based co-ordination mechanism governs the interaction, setting out specific and embedded rules for the interaction. Besides the organisational principles of him accepting a certain quota of computer-cab jobs monthly, the specific co-ordination of these jobs is carried out according to a pre-programmed set of steps where he selects one of the available jobs on the computer cab terminal. This is significantly different from the use of mobile phones, which serves either as a resource for reflexive action or as a means for external others to request attention.

Driving a taxicab around London waiting for customers to hail the cab, receiving jobs through the computer-cab system or positioning the cab in a taxi rank is essentially mobile work responding to emerging and greatly varying customer needs across the city, at different times of the day, days in the week or seasons. However, the work itself is far from being entirely governed by this emerging demand. Ray's work is a continual balance of reconciling the emerging situation with his extensive knowledge and routines combining embrained, embodied and embedded knowledge (Blackler, 1995). The years of driving around the streets of London on a scooter in order to pass the strict exam shaped a generative process between the *knowledge* of streets and places of interest and the *knowing* shaped by the process of driving and experiencing them (Cook and Brown, 1999). This provides a theoretical grounding, which for the subsequent years is related to experiences of shifting demand for taxi work, emerging routines and other constraints. Taxi drivers are armed with extensive knowledge and routines to help them engage in balancing the paradox of emerging actions and planned intervention on a daily basis. In terms of the very real paradox of picking up customers in the street and by doing so possibly endangering their own safety, taxi drivers will in response adopt micro-routines to best counter the most prominent specific risks, for example, driving past a hailing customer to provide a little more time to assess the likelihood of danger (Gambetta and Hamill, 2005).

Mobile services transcending the boundary between the cab driver and the surrounding world not only influence the cabbies' ability to obtain fresh information and socialise from within the cockpit but also alter the relationship across the glass wall separating the driver from the customer. When a customer enters the cab, it becomes a place of doing business and therefore represents a place of 'frontstage' behaviour where the driver represents a 'favourable definition' of the service offered (Goffman, 1959, p. 83). Drivers socialising at taxi ranks or in a cabman's shelter are more likely to display 'backstage' behaviour where participants can more openly discuss issues and display casual behaviour. The mobile phone can make the cab into an extension of the client's office or home but can also allow the cab driver to extend socialising and display some characteristics of backstage behaviour whilst carrying a customer. Due to the physical separation of the driver and the customer, it is possible for both parties to flexibly negotiate to what extent the co-presence requires interaction or to what extent the glass divider represents a separation of the cab into two independent cells. Technically Ray has the upper hand, as he controls the microphone transmitting any conversation across the glass barrier. Ray screens calls to his mobile phone and sends an SMS message to the caller simply stating 'POB' or picks up the call and merely states this. It is a coded explanation that he has a 'passenger on board'.

Mobile workers actively shape the places they encounter in order to be able to conduct their work, such as the practice of stalling, i.e., commuters making claim to a bounded space (Goffman, 1971; Felstead *et al.*, 2005, p. 22). An example of this can frequently be observed on commuter trains, where tray tables are folded down or briefcases are placed on seats to ensure the commuter can occupy two seats. Commuters may engage in 'planful opportunism' (Perry *et al.*, 2001) by preparing for the unpredictable through installing as much background information and as many reference works as possible on their notebook computer. Axtell *et al.* (2008) discuss the relationship between the affordances of the context influencing the work activities chosen by commuters, for example, establishing routines of which seat to choose on the commuter train. Within the context of the Black Cab, both cab drivers and passengers can use the mobile phone as means of creating a boundary for interaction by becoming absent present (Gergen, 2002) talking on their mobile phone. This *virtual stalling* creates a barrier for interaction across the glass divider.

4.5   Case 3: Khalid – foreign exchange trader

Vignette 3

Khalid is having dinner with his family in a restaurant somewhere in the Middle East. He works as a trader for a large Arabian bank and is part of a small group of foreign exchange traders that extend the banks' trading hours throughout evenings and nights equipped with a trading pager, a mobile phone and a PDA with web-based trading services. Khalid takes a short break from the discussion of what his children have done in school that day as he checks his trading pager, a Reuters SmartWatch, to see if any rate changes at the New York Exchange are influencing his positions. The small group of traders entrusted to perform off-premises trading negotiated their positions before leaving the trading floor earlier in the afternoon. This provided a common understanding of their individual trading limits. Khalid will only occasionally need to call fellow traders in order for them to co-ordinate between themselves that the collectively arranged trading limits are not exceeded. When he has completed a trade, Khalid uses his mobile phone to record the trade onto a telephone answering machine at the bank for further back-office processing the following morning (study 3 in Table 1.1).

Context 3

A large portion of foreign exchange is traded over the counter in markets that operate round the clock, i.e., outside the regulated exchanges (Kane, 2003). There are no geographical monopolies in the over-the-counter foreign exchange market. Between 50 per cent and 75 per cent of daily turnover is cross-border during the business hours of the financial centres of London, New York and Tokyo. This suggests that one side of many transactions occurs outside of the regular working hours for many financial institutions that have operations around the world (Kodres, 1996). The geographical spread of foreign exchange has created a challenge for financial institutions in terms of coping with the 24-hour nature of the market. The development of global networks and mobile services has greatly supported the demand for constant trading. Knorr-Cetina and Bruegger (2002, p. 163) point out that traders continuously keep track of the market through Reuters' hand-held screens and satellite TV channels after closing hours. They suggest that for traders 'the most fascinating part of their environment is the market – with which they appear to be excessively engaged not only during working hours but also during evenings and weekends'. Schwager (1992, p. 60) argues that traders often articulate their intense and 24-hour engagement with markets in interviews and conversations, and exemplify this with a quote from a trader stating 'you work, relax, eat, and literally sleep with the markets'.

4.6   Mobile trade-offs

The Middle-Eastern bank where Khalid works has conducted off-premises trading since 1997 using a variety of mobile services. Trading during the day in the trading room is a tightly co-ordinated effort where trading is conducted through computer screens and telephone conversations, and the traders 'observe the market and work on it' rather than work in a traditional trading pit where traders 'live the market with their bodies and voices' (Zaloom, 2006, p. 141). Mobile trading outside of normal trading hours is only conducted by a select few specially authorised and trusted traders. The trader obtains market information through a Reuters SmartWatch trading pager, a Pocket PC with Reuters foreign exchange web services installed and additionally from financial SMS services on his or her mobile phone. The Reuters SmartWatch trading pager monitors specific markets and currency exchange rate fluctuations, and supports the trader in accessing financial information whilst on the move. It displays foreign exchange and money market rates, derivatives, capital markets, stocks and indices. It can also be customised for local and regional information. The pager has a database of over 500 instruments and is designed to chart the progress of selected financial markets. The SmartWatch also provides technical analysis criteria that can be tailored to the requirements of each individual trader. It has charting pages supporting traders in creating ten customised market trend charts.

The SmartWatch 'Limit Alerts' function allows the trader to create trading limits and alerts in order to be notified instantly when these limits are breached. The pager can also be set to alert the trader if any of the prices from the most recently viewed display screen change. The news page of the SmartWatch displays the latest financial news headlines and the pager can be connected to a personal computer for more powerful and flexible features. Some of the traders are also using the more advanced service offered by Reuters foreign exchange web services running on a networked PDA. The Reuters Middle East office chose the Middle-Eastern bank from case 3 to pilot this suite of services in the region. The PDA version provides similar services to the SmartWatch on a more powerful technological platform. Furthermore, the PDA supports mobile transactional services with specific leading banking institutions. The user-defined alerts, limits and views constitute the affordance of increased intimacy through configurability and the alerts constitute self-instantiated planned performances.

Deals are mostly agreed on the mobile phone. The bank must adhere to requirements set by the financial authorities, for example, managing the potential risks associated with the use of mobile services. This implies the requirement of traders to record their transactions by dialling a secure and protected recording system within the bank. Using their mobile phone, traders call a telephone number protected by secure codes and record the transaction immediately once the deal is agreed with the counterpart. The archived records contain information about amounts, currencies, counterparts and any additional necessary remarks. The back-office internal control department transcribes these records on a daily basis and enters them into the bank's back-office system. This is later followed by a daily written report by the trader. The mobile phone is also used for contacting clients and other traders in case-specific transactions that need to be negotiated or for co-ordinating collective trading limits with fellow traders. Traders obtain market information on their mobile phones either through calls to colleagues, SMS messages received from banking information services or SMS messages exchanged with other traders.

Engaging in off-premises trading places demands on the traders' family life. This is one of the main reasons for maintaining light-touch management control during off-premises trading and traders are generally left to themselves. However, the constant demand for availability can lead to problematic situations if traders do not at times engage in minimal co-ordination with fellow off-premises traders to negotiate trading limits. For example, one trader stated: 'When I look at my Reuters SmartWatch while I am with my friends, they know I have work obligations. They do not mind that I stop talking, and start using my phone to make deals' (Al-Taitoon, 2005, Trader #5, p. 145).

Another trader highlights time management and interruptions as a contentious family issue: 'My wife got used to the alerts that I receive from Reuters SmartWatch at midnight. She does not mind that anymore. We had problems in the beginning [...] The issue is sometimes with the children who you should give the time they need [...]; therefore, I sometime explain to them what I am doing' (Al-Taitoon, 2005, Trader #7, p. 144). Several traders emphasise the dual pressures of family life and continuous trading and the overarching need to keep connected to the market no matter what. A third trader argues that he struggles 'when the market moves against me, and if at the same time I have problems at home' (Al-Taitoon, 2005, Trader #9, p. 143). Traders express the importance of continuing to trade irrespective of their family situation.

The challenges to mobile users' boundaries between home life and working life is well documented, for example, by Gant and Kielser (2001); Ahuja *et al.*'s (2007) study of factors affecting mobile worker turnover; Wajcman *et al.*'s (2009) survey of Australian mobile phone practices; Ling's (2008) discussion of interaction with his plumber; and Cousins and Robey's (2005) analysis of mobile workers' boundary management. The problems of clearly separating working life and home life is not one exclusive to the age of mobile services. The intensification of work increased reliance on computer-mediated information and interaction management led to increased needs for the explicit management of work-life boundaries (Ashforth *et al.*, 2000; Golden and Geisler, 2007; Cousins and Varshney, 2009).

For traders such as Khalid, the challenge consists of balancing a range of conflicting technology performances in order to negotiate the organisational need for emergent decision-making and for planned intervention. The interdependencies during the day are replaced with explicit attempts to render traders' work independent out of hours. However, the need to document transactions within the organisation and the occasional need to negotiate trading limits still presents some interdependencies requiring specific technology performances. This can be synthesised as the trader acting alone yet being accountable to the organisation, and the primary mechanism for ensuring this tension is the organisational process of carefully selecting traders entrusted with this independence. Khalid and his mobile trading colleagues are also required to carefully and independently negotiate being constantly connected to and yet selectively disconnected from the market.

4.7   Anytime and anywhere, anyone?

Hiro the Tokyo CEO, Ray the London taxi driver and Khalid the mobile trader all engage extensively in interaction with remote clients, colleagues and systems mediated by mobile services. Hiro has cultivated technology performances allowing him to roam the streets of Tokyo gaining inspiration between meetings. Ray and his colleagues have as a profession roamed the streets of London for more than 400 years, but with contemporary mobile services they have become aware of important changes beyond their immediate location, responding in a more co-ordinated manner and at the same time keeping in touch with individuals who really matter to them. Khalid uses mobile services to continue trading currencies on global markets beyond his normal day's work in the trading room.

Given these observations of these individuals' balancing of emerging and planned technology performances, it is tempting to employ the much-used assumption that one of the consequences of mobile services is the ability to conduct work *anytime, anywhere*. The technological promise of instant and ubiquitous interaction has greatly excited the imagination of marketing people, technology vendors and researchers alike, leading to the frequently used notion of *anytime, anywhere* interaction (Kleinrock, 1996).[11] The emphasis on *anytime, anywhere* access to information and people goes beyond discussions of mobile services. It relates more broadly to the emergence of global Internet infrastructures often captured by Cairncross' (1997) powerful meme pronouncing 'the death of distance'. This relates to the declining costs of communication and the increased reliance on electronic communication. Malone (2004), for example, argues that such development enables the decentralised and distributed organisation of work. The idea of simultaneously interacting with others whilst mobile has inspired highly optimistic accounts of technological possibilities, such as Davis' (2002) exploration of the future of knowledge work, Kleinrock's (1996) discussion of infrastructure requirements and Makimoto and Manner's (1997) early exposition of the 'digital nomad' who engages unhindered in fairly unproblematic global nomadic working.

Much of the technologically inspired debate has assimilated the simple assumption of assigning both the promises and realities of mobile services to the shorthand of 'anytime, anywhere interaction'. This has resulted in a common assumption that the organisational consequences of the technology are primarily flexible working arrangements where work can be conducted outside of traditional physical constraints. The simple linking of interactivity at any time conducted anywhere leading to flexible working arrangements where work is then accomplished free of traditional constraints is an alluring image of progress, freedom and flexibility. However, it is a very simple image that only reveals a small part of a complex relationship between technological affordances and human actions. The issues are more complex than the initial technology-focused accounts of how opportunities alter behaviour in a linear manner, and these accounts tend to neglect the socially and organisationally situated nature of interaction (Wiberg and Ljungberg, 2001; Pica and Kakihara, 2003). Not all moments in time and place are equally suitable for all conversations and neither are all situations (Ljungberg and Sørensen, 2000; Kakihara *et al.*, 2005). So, while the opportunities for interaction may be ever-present, so are constraints, preferences and everything else that inform the individual disposition to interact (Ciborra, 2006). However, while the opportunities to connect with remote people and services may remain stable over time as instant opportunities, individual predispositions can rapidly change. Arnold (2003) captures the tensions between the technological affordance of instant connectivity and the richness of the context as the Janus face of technology in action.

The situating constraints of work can emerge from a variety of sources. The Tokyo CEO, the London Cab driver and the Middle-Eastern trader are generally not constrained by the organisation or by colleagues, but rather by themselves or by the work itself. Hiro is the owner and the CEO of a small organisation and will therefore have considerable discretion. Ray owns his own cab and equally will be able to make most of the decisions regarding the organisation of his work all by himself. For Khalid, trading out of hours is a careful balance between highly independent mobile working and being a father to his family. The practical constraints of work can influence times and places of work (Wiberg and Ljungberg, 2001). When a Black Cab driver's extensive practical knowledge of how to engage with the topology of London to optimise income meets emerging customer demands, traffic situations, etc., the driver's technology performances bind together mobile services and the specifics of the situation. Explaining this in terms of 'anytime, anywhere' is both misleading and simplifies the underlying complexity of the socio-technical relationship. Taxi drivers wait or drive around for potential customers in places they are likely to appear, photocopy repair engineers need to travel to where the customers keep their equipment (Orr, 1996), telephone repair engineers can only fix some of the problems through remote access (Wiberg and Ljungberg, 2001), ship insurance inspectors can only inspect a tanker from inside the ship (Kristoffersen and Ljungberg, 1999) and plumbers do their plumbing where the pipes, taps and sinks are situated (Ling, 2008).

Hiro, the Tokyo CEO, may be a digital nomad with extensive individual discretion, yet it is not the case that his working life is 'anytime, anywhere'. Physical meetings form an essential part of his everyday activities. Furthermore, his extensive discretion to select and shape his workscapes (Felstead *et al.*, 2005) translates into specific choices for the purposes both of collecting inspiration and managing the organisation.

4.8   The unbearable lightness of situations

Co-present social interaction is characterised by a rich set of rituals establishing the interaction, managing turn-taking amongst participants and repairing it when problems arise (Goffman, 1959; Collins, 2004). The fluid turn-taking between individuals engaging in face-to-face conversation is 'locally managed, party-administered, interactionally controlled, and sensitive to recipient design' (Sacks *et al.*, 1974).

Social interaction mediated by mobile services utilises existing practices and can both reproduce existing rituals and play a role in emerging rituals (Ling, 2008), for example, through new interaction practices such as connected presence where frequent mediated interaction merges with co-present interaction to form a whole (Licoppe, 2004). The transition between situated and mediated interaction necessitates additional effort to both establish the interaction and integrate the co-present and the remote, as well as to repair this when problems occur – for example, gestures signalling to present others that a mobile phone call is important and that it needs to be answered.

For Hiro, channelling synchronous phone calls into asynchronous emails facilitates this ongoing negotiation between the co-present and the remote. Remote requests are buffered and filtered, supporting a fluid process of continuously evaluating whether a request should be allowed to interrupt a given situation. Like other professionals dealing with a range of audible cues, Hiro has trained himself to ignore information deemed unimportant and to instantly react to that deemed important based on the different noises his mobile phone emits generated by the various filters. Similar examples of the importance of subtle cues establishing peripheral awareness have been documented, for example, in trading rooms (Heath and Luff, 2000).

Flexible working arrangements, such as flex-working, mobile working, home working, hot-desking or sharing offices, are often associated in the press with employee benefits, allowing work to be conducted from the garden, the beach or by a hotel pool, as illustrated in the *Washington Post* article on digital nomads (Rosenwald, 2009). Such emphasis on promises of fluid working practices emphasises technological promises and jointly the two weave a glossy tapestry of opportunities, as discussed in the previous section on anytime, anywhere interaction. However, the ability to actively choose particular contexts of interaction whilst avoiding abandoning connectivity to essential resources and contacts changes human mobility from an employee benefit to an issue of effective decisions and, as such, makes it an organisational priority.

Brown and O'Hara (2003, p. 9) argue that the main reason for the geographical mobility of the workers they study is to meet different people. The active engagement with others through the act of placing oneself in their company becomes paradoxically significant the easier instant remote connections can be made. Actively choosing a context when constraints do not impose it is an important individual statement that 'I have explicitly chosen to be with you as opposed to all the other places I could have been'. In the duality between individuals being present to some and remote to others, the co-existence of presence and remoteness (Licoppe, 2004) implies that the choice of context can to a greater extent be viewed as a strategic rather than an incidental choice. Such choices are informed by both planned interventions and emerging actions.

Not all situations offer equal access to the resources needed to engage in the necessary information working. The paperless office is far from a reality, although technological developments reshape the role of paper (Sellen and Harper, 2003). In terms of micro-mobility (Luff and Heath, 1998), the affordances of, for example, paper-based records, notes and ledgers make it a stable part of most organisations and for mobile workers this implies that all necessary resources may not be available in all situations (Sherry and Salvador, 2001). With increased computerisation, the necessary documents may perhaps be available electronically, and the role of paper may change from the master and reference resource to a disposable copy. When the mobile worker is able to receive or carry all the necessary resources for his or her emerging decision processes, the circumstances may not be conducive for organising the wealth of information provided. The technology is paradoxically devised, aiming at increasing the ability to provide remote access to the necessary information and contacts, and yet, once this promise is fulfilled, the mobile worker may not find himself or herself able to deal with the information or connectivity sufficiently (Sherry and Salvador, 2001).

Characteristic for the three cases discussed so far is the careful cultivation of the relationship between the information available and required. For Khalid, the challenge is being able to interpret and filter all available information presented on a small LCD screen while away from the trading room with its dedicated extensive support for complex information management. Therefore, he spends considerable time honing the information flows, alerts and filters in his Reuters SmartWatch. Hiro also has limited access to complex information when receiving a stream of information and interaction requests through email to his mobile phone. He spends significant effort on both enticing customers and colleagues to feed everything through mobile email and on then honing the filters and alerts to help him manage the complexity. For Ray, 'the Knowledge' represents a sophisticated process of accommodating the instant need for complex topological information by cab drivers.

Perry *et al.* (2001) argue that mobile workers opportunistically plan for a range of possibilities to engage in work prior to travelling – planful opportunism. They prepare for the unpredictable by installing as much background information and as many reference works as possible on notebook computers. This relates to the classification of mobile information technology support in pre-, in- and post-mobility activities (Al-Taitoon and Sørensen, 2004).

4.9   Rhythms of interaction

This section explores *rhythms of interaction* as alternations in coupling and decoupling with technology (Sørensen and Pica, 2005). Rhythms of interaction can be constituted by brief recurring utterances of turn-taking in conversations or by engagement with mobile devices and remote services. In short, it is a measure of the shifting attention.

Lefebvre (2004, p. 15) argues that 'everywhere where there is interaction between a place, a time and an expenditure of energy, there is rhythm'. He further characterises such rhythms in terms of: 1) repetition, for example, relating to gestures, movement, action and differences; 2) interferences of linear and cyclical processes; and 3) phases of birth, growth, peak, decline and end.

The notion of rhythms in social life has been investigated by a number of researchers: Zerubavel (1981) characterises social life by a multitude of temporal arrangements and rhythms – for example, schedules, calendars and recurring events; Licoppe and Smoreda (2006) explore rhythms of social interaction; Fortunati (2002) characterises social relations in terms of rhythms of absence and presence; and Lefebvre (2004) provides a philosophical account on rhythms in everyday life. Rhythms can signify the alternation in the intensity of being busy (Lee and Liebenau, 2002) and the juggling between virtual and situated spaces of interaction as they unfold in the action (Bunzel, 2002).

Green (2002) proposes three interconnected domains of: 1) rhythms of use of mobile services – the duration and sequence of interaction between the user and the mobile device; 2) rhythms of integrating the use of mobile services into everyday life – the embedding of the use of mobile services in social relationships; and 3) rhythms of the use of mobile services in relation to institutional change – infrastructural elements enabling the use of mobile services.

Hiro uses his mobile phone to check his email before placing the phone back in the pocket of his jacket. Equally, Khalid engages and disengages from the fluctuating currency market through his SmartWatch during evenings when he regularly checks currency movements. Both of them engage in continual rhythms of coupling and decoupling, forming contextual resolutions of emergent and planned performances. Coupling and decoupling mobile technologies are pre-cognitive resolves of routines and improvisation in the 'now' (Weick, 1998; Ciborra, 2002). Rhythms reflect the meeting of improvisation and routines, emerging self-determined needs, situational circumstances and stipulated procedures. For Khalid, the inclination to check the market, the alerts triggered by market fluctuations, the organisationally stipulated procedure for recording trades and a range of additional aspects all merge into specific decisions to engage with the portfolio of mobile services at his disposal. A completed foreign exchange trade triggers the requirement for him to document it.

Longer rhythms relate to patterns of technology performances in recurring work situations – for example, when the cab driver Ray starts and stops the taxi meter at the beginning and end of each fare, or throughout the day enters his location into the cab computer to let the dispatch office know what part of London he is in. Seasonal rhythms of work can also influence technology performances. During the Christmas period there is much higher demand for cabs than at other times of the year. Here the cab drivers rely less on the centralised computer dispatch system, as there is plenty of work. However, during January, February and August, when demand is greatly reduced and even experienced cab drivers can struggle to find enough work, the drivers rely much more on computer-cab systems as a source of income (Elaluf-Calderwood, 2008, p. 148).

4.10   Cultivating fluidity

Techno-optimistic predictions of boundary-free mobile working, as discussed previously, emphasise friction-free technological opportunities. The ideal of individual mobile workers experiencing their activities and interactions merge into *fluid working*, where a purposeful stream of decisions, actions, information and interactions all merge into a desirable whole for the individual, is of course merely an ideal. The reality is often one characterised by significant second-order activities managing the interaction with others (Nardi and Whittaker, 2000; Kakihara *et al.*, 2005; Wiberg and Whittaker, 2005). Fluid working perhaps best relate to Csikszentmihalyi's (2002) psychological account of achieving a mental state of complete immersion through combining clear goals, an altered sense of time, concentration, a sense of control, a loss of self-consciousness, balancing challenge and ability, direct feedback, action awareness, and effortlessness of the actions. Such a flow is more likely to be achievable when the individual is engaged in isolated solitary confined activities, where the environment is entirely within the individual's control, than in situations where the outside world of people and services place demands for attention and the need to be consulted in order to accomplish the work.

Individual processes of coupling and decoupling mobile services relate intimately to situated interaction – for example, face-to-face meetings or the use of various other instruments and technologies. Rhythms of interaction with mobile services will to varying degrees take precedence or be subservient. Generally, there is a complex and possibly uneasy relationship between situated and mediated interaction (Ling, 2008). Mobile interaction can transcend local and remote context barriers and as a result cause interruptions. Most people have probably experienced situations where mobile phone calls immediately take precedence in the midst of face-to-face conversations, or cases of absent presence (Gergen, 2002) where a person is physically near but is entirely absorbed by the activity of reading a message on his or her mobile phone. While such acts can merely be civil inattention (Goffman, 1959), they can also take place when attention is assumed.

With years of internalised knowledge of the streets of London and the shifting need for cab services, Ray may indeed be able to create such an environment of flow where there is a perfect combination of his highly advanced skill level and the challenges he experiences in his daily work.

Khalid may also experience situations where off-premises trading sessions are solitary battles against the marketplace conducted in a state of fluid activities. However, the sense of control may indeed be as misplaced as a surfer feeling entirely in control of his or her actions when surfing large swells. Uncertainty, interactive complexity and tightly coupled elements make the global financial market vulnerable to what Perrow (1984) calls 'normal accidents' and, as such, stable situations of fluid trading may be difficult to maintain.

In the case of Hiro, his efforts to channel all his interaction into mobile email can be seen as an attempt to gain slightly more control by retaining the initiative to respond, prioritise and ignore interaction. The extensive use of filters equally affords him some form of control and awareness of when a person deemed important contacts him. Whilst it may be conceivable that Hiro, through his extensive secondary efforts to streamline and arrange his interaction, will be able to achieve an increased sense of fluid interaction, it is not likely that this will easily enable the sense of flow as characterised by Csikszentmihalyi (2002).

According to Mintzberg, executives play ten basic roles within three broad groupings:

> The interpersonal roles describe the manager as figurehead, external liaison, and leader; the information processing roles describe the manager as the nerve center of his organization's information system; and the decision-making roles suggest that the manager is at the heart of the system by which organizational resource allocation, improvement, and disturbance decisions are made. (Mintzberg, 1971, p. B-97)

Such contexts for decision-making are inherently characterised by a reliance on intense social interaction (Kurke and Aldrich, 1983), as Mintzberg argues:

> Because of the huge burden of responsibility for the operation of these systems, the manager is called upon to perform his work at an unrelenting pace, work that is characterized by variety, discontinuity and brevity. Managers come to prefer issues that are current, specific, and ad hoc, and that are presented in verbal form. (1971, p. B-97)

Hiro cannot shield himself from the uncertainty, variety and discontinuity of interaction patterns, but he engages in complex processes of managing these interaction patterns to optimise his own and the firm's performance in contextual ambidexterity with concern for both short-term exploitation and long-term exploration.

There is no guarantee that concerted attempts by a mobile worker to engage a diversity of mobile services will result in fluid performances. Interruptions form a key concern for the understanding of enterprise mobility. The distractions of continuously checking email are examples of self-imposed interruptions in order to read or send emails, but distractions can also be directly instigated by others or by various systems and services.

4.11   Managing interruptions

According to McFarlane and Latorella (2002, p. 19), interruptions can be characterised in terms of: 1) the source of interruption, for example, oneself, another person or a service; 2) the individual characteristics of the person interrupted, for example, cognitive ability and goals; 3) the method of co-ordination, for example, no co-ordination or scheduled co-ordination; 4) the meaning of the interruption, for example, accidental, an alert or a reminder; 5) the method of expression, for example, verbal or facial; 6) the channel – face-to-face, mediated by a direct channel or by a machine; 7) changed human activity – individual or collective; and 8) the effect of the interruption, for example, changed activity, loss of focus or changed awareness.

There is a significant amount of research on interruptions, in particular within psychology and HCI. The website www.interruptions.net contains a large collection of research resources. McFarlane and Latorella (2002) forward a taxonomy of human interruptions; McFarlane (2002) discusses the implications for design; Dindia (1987) explores the role of gender in interruptions; and Wiberg and Whittaker (2005) relate interruptions and availability management. In addition, several researchers have discussed interruptions at work (for example, Rouncefield *et al.*, 1994; O'Conaill and Frohlich, 1995).

For Ray, fluid mobile working consists of constant physical interruptions with the yellow light on the roof of the taxicab clearly indicating to a potential customer if he can be interrupted or not. Interruptions from mobile phone calls can be flexibly managed in a variety of ways, as regulations require that all cab drivers use hands-free connections to mobile phones. The POB code signifies to insiders that discussions should be brief.

Hiro's extensive use of filters and a variety of sound notifications for the incoming requests for his attention on his mobile phone exemplifies an attempt to manage interruptions. It is in essence a highly advanced and more diversified version of the common practice to put mobile phones on silent in meetings or at presentations to avoid disruption.

Khalid deploys a variety of profiles and filters serving as a boundary object (Star and Griesemer, 1989) between him and the ever-changing global foreign exchange market. In this way, carefully crafted informational events are planned to occur when the criteria defined by the trigger are fulfilled. These events may or may not constitute interruptions, depending on the situation. If Khalid is actively gazing at market fluctuations, the events can serve as further information, but if he is asleep, it can constitute an interruption.

As a means of managing the rhythms of interaction, interruptions can have beneficial effects. Ljungberg (1997) discusses the example of Karen, a clinical trial manager in a drugs company who, before a business trip, actively sought interruptions from colleagues by sending them all an email to please drop by her office if there was anything they wished to discuss. In this particular context interruptions represent the clearing of any necessary interactions prior to her being out of the office for a few days. Some interaction can be perceived as beneficial even if it is deemed disruptive (O'Conaill and Frohlich, 1995).

The widespread use of mobile phones and other dedicated devices for mobile email use is a good example of such forceful change in mobile information work. The phenomenon of constant email connectivity through mobile devices is frequently associated with notions of enslavement and misuse, and devices are referred to as 'crackberries', alluding to the addictiveness to the technology caused by the ability to constantly check email. Academics and the press both use the term, for example, informing a wider audience of this problem (*The Economist*, 2005), in research of investment bankers' use of mobile email (Mazmanian *et al.*, 2005) or relating to medical advice on how to determine the level of addiction (Taylor, 2007).[12] *The Economist* (2005) points out that such addiction to technology is nothing new: 'This has been going on since the mid-19th century, when telegrams were introduced. "The businessman of the present day must be continually on the jump – he must use the telegraph," grumbled one New York merchant in 1868.' This relates directly to Mick and Fournier's (1998) paradox of freedom and enslavement as an inherent paradox of technology.

For the 'hyperconnected' (Aducci *et al.*, 2008) mobile worker constantly being offered the opportunity of checking his or her email wherever he or she may be, this is a distraction that may be too difficult to ignore. There is no barrier between intention and action due to the intimacy of the relationship between the service and the user. The lure of the easy activity allows the individuals to scout for a more interesting world outside of a dull meeting by escaping into the world of unexpected emails. Here micro-decisions constantly weigh up if the present is more or less interesting than the unknown possibilities of unread emails. Mobile workers checking their email clients repetitively, again and again and again, are fixed in a trance like commuters waiting on the platform for the announcement of what track their train will depart from and not being able to take the eyes off the information board. The possibilities of anticipated and unanticipated emails both offer an alluring distraction of satisfying curiosity and of feeling in touch with current activities. Rapid responses to emails can play an active part in ongoing interpersonal rituals by signalling to the sender that acute attention is given to them and will hence help to improve his or her perception of the person offering rapid response (Goffman, 1959; Ling, 2008).

Artificially created barriers between intention and action can be a necessity in order to enforce a re-balancing of interruptions and deeper engagement – for example, only reading and responding to email in the afternoon or physically removing the mobile phone. Such technology performances serve the purpose of re-balancing the tension between fluid interaction across a range of participants and areas of interest towards creating barriers encapsulating the essential subjects and participants. Jackson (2008) provides an extensive discussion of distractions, arguing for the erosion of attention in modern society and for the need to re-focus and prioritise as a wealth of opportunities for distraction emerges. Carr (2010) similarly argues that constant connectivity to vast repositories of information can both greatly inform use but also provide endless distractions and diversions from concentrated insights.

4.12   Mobile overload

The subject of interruptions relates to information and interaction overload, which can be seen as unintended consequences of computerising information management (Hiltz and Turoff, 1985; Mathiassen and Sørensen, 2008). Unintended consequences of mobile services are likely to emerge if assumptions behind the design do not fit the realities of use (Markus and Robey, 2004).

Generally, information and interaction overload characterises the last stage in the diffusion process that starts with the perception of the potential benefits of adopting a specific information technology, followed by the diffusion of the service and leading to it becoming an integral part of everyday life. The service may then be used so widely that its consequences are different from those intended.

Traditionally, the focus has been on information overload (Hiltz and Turoff, 1985; Mackay, 1988). Ljungberg and Sørensen (2000) contrast information overload with interaction overload in terms of the former emphasising the individual ability to process information, with the latter emphasising interactional preferences. Mathiassen and Sørensen (2008) present a conceptual framework characterising the differences between information and interaction overload in terms of the underlying information services applied.

Research has explored information and interaction overload from a range of perspectives. Jacoby (1984) and Malhotra (1984) study information overload in the context of marketing to consumers. Eppler and Mengis (2004) review the literature across a range of academic fields. Iastrebova (2006) explores coping strategies applied by managers experiencing information overload. Mackay (1988 and 2000) studies how different personality types deal differently with information overload and how users and technology co-adapt to manage overload. Hiltz and Turoff (1985) and Maes (1994) discuss how to design information systems to avoid information overload. Wiberg and Whittaker (2005) discuss cognitive overload and interruptions in the management of availability.

The notion that overload is a universally measurable issue will be highly misplaced. Firstly, unintended consequences are influenced by the adaptive behaviour of the involved actors. They can then generate coping strategies that reduce the problem over a period of time (Schultze and Vandenbosch, 1998). Secondly, overload is both individually and contextually conditioned in the same manner as information richness. Defined in the traditional sense, information richness is a property of a medium (Daft and Lengel, 1986). However, a broader contextualised view reveals an alternative understanding of information richness as shaped by the context of interaction (Ngwenyama and Lee, 1997). What might be perceived as a misfit in one situation, may, in another situation or by another actor, be perceived as normal and useful (Mackay, 1988). The understanding of information overload must be informed by the particular context in which the interaction takes place.

4.13   Summary

The role of technology performances for managing the tension between fluidity and boundaries has in this chapter been discussed in terms of the cultivation of both individual fluid interaction and the barriers shielding the individual mobile worker. This raises a number of issues for discussion – for example, the problematic one-sided assumption of anytime, anywhere interaction emphasising fluidity over a continuous process of balancing fluidity and boundaries. The role of explicitly selecting the physical context for interaction exactly emphasises the orchestration of planned fluidity through fixing boundaries. Purposeful or incidental shifts in the coupling and decoupling with mobile services are discussed as rhythms of interaction. The cultivation of fluid working can be related to the notion of flow, which indicates specific assumptions about the degree of control over, and the character of, interruptions. This was followed by the discussion of interruptions as an integral part of managing the fluidity of and barriers for individual interaction. The issue of interruptions was then related to information and interaction overload.

These challenges were related to the three cases of a CEO, a taxicab driver and a mobile foreign exchange trader, whose use of mobile services signified ongoing attempts to establish flow in their working practices through the dual cultivation of interaction fluidity and boundaries. The following chapter considers the role of technology performances for mobile workers' cultivation of fluid collaboration and barriers for collaboration. Here the emphasis is on individual and collective practices aimed at facilitating the negotiation and resolution of mutual interdependencies.

5
Collaboration – Transparent Interdependencies

This chapter explores mobile technology performances aimed at supporting collaboration. Here distributed and mobile workers engage in co-ordinating mutual interdependencies with the purpose of resolving these and as a result producing co-ordinated outcomes. Specifically, the chapter draws out themes of relevance to the transparent negotiation of interdependencies through collaborative fluidity and boundaries for interaction.

This chapter initially presents and discusses four cases: 1) two UK response vehicle police officers engaging with colleagues and others using a complex portfolio of mobile services; 2) a security guard supported by mobile services relying on RFID tags embedded in his work environment; 3) an industrial waste management lorry driver, also using RFID technology; and 4) a Japanese town planner collaborating using a mobile phone and a notebook computer.

5.1   Case 4: John and Mary – police officers

Vignette 4

In the South of England two police officers, John and Mary, are getting ready to begin their shift driving a response vehicle. They have just attended the morning briefing where yesterday's incidents, as well as ongoing issues, are discussed. They have just compiled outstanding paperwork from the day before and checked their email on a stationary computer, when a domestic disturbance incident is announced on the police radio transmitting incidents to the office. They quickly equip themselves with shoulder-mounted personal radios and stab-proof vests and get into the assigned response vehicle among those parked at the back of the police station. They drive at high speed with sirens and blue lights in full action through a small town and down country lanes, with John driving the car and Mary sitting in the passenger seat. Whilst driving to the incident they are engaged in two important tasks: one is to ensure that they arrive as fast and as safely as possible and the other is that they have as much information as possible about the incident they will very soon be attending. Mary asks the control room for details about the incident. A few seconds later the in-vehicle Mobile Data Terminal (MDT) beeps and Mary reads out loud the information provided by the control room. She tells John that a man allegedly involved in the incident in question has a history of domestic violence and that he is considered dangerous.[13] He tried to attack a police officer during a previous incident somewhere else. The information enables the officers to discuss the situation ahead in order to assess what risks may be involved and how to prioritise their efforts. John suggests that Mary asks the control room for back-up in case other cars are in the area, which she does through her radio. He also asks her to see if the log describing the incident mentions other incidents at that particular address – it does not. The log contains information about a woman in a state of distress at the address and that the control room has organised an ambulance to meet the police officers there. Throughout the high-speed journey towards the incident, Mary guides John in navigating intersections by giving clear signals at her side of the road.

The response vehicle arrives at the incident situated on a council estate – local authority-owned rented housing. A small group of people are standing outside the building. John starts making his way to the flat. Mary uses her mobile phone to call the victim's mobile phone in order to get an update on the status of the incident, but the call is not answered. She uses her personal radio to frequently update the control room about her and John's location as well as the general status of the incident. They decide to go to the third floor, where the flat is located. On arriving at the front door of the flat, they see that the ambulance and another response vehicle have just arrived. John knocks on the front door of the flat repeatedly, but no one is answering. The neighbour comes out of her flat and states that she was the one making the initial call to the police and that the female victim is still inside the flat. She also tells John and Mary that the male perpetrator has now run away. While the neighbour provides further details, the front door slowly opens. Mary uses the radio to update the control room of the situation and a woman emerges, clearly in a state of shock, bleeding from the nose. Mary asks if anyone else is inside the flat. The victim answers that there is nobody and that the perpetrator escaped using her car. John cautiously makes his way into the apartment and verifies that there are no more people inside. Mary gives the paramedics permission to enter the scene after it is now clear of any possible dangers. While the victim is in the care of the paramedics, Mary asks her some questions regarding the incident. The victim provides the registration details of her car and the name of the perpetrator. Mary puts out a warrant on the vehicle registration number through her shoulder-mounted radio and asks the victim to describe the perpetrator and how he is dressed. Mary keeps updating the control room with details as well as taking notes in her police notebook. She attempts to calm the woman down and reassures her that the police force will take care of the situation. John and Mary then get in the car and consult the MDT and the in-vehicle radio for further incidents to attend. Towards the end of the shift, after a few more incidents, they slowly drive back to the police station. The MDT is turned off and the in-vehicle radio keeps broadcasting incidents. Back in the office, they document the incidents attended during their shift. They wait for an extensive period of time on the phone to receive a unique identification number for each incident from a central police service and complain bitterly about this waste of time. They are, however, satisfied that they did not have to arrest anyone during their shift, as the associated paperwork and activities of booking someone into custody would have taken most of the shift. They generally feel that time spent in the office or at the arrest takes them away from the job they signed up for and like (study 4 in Table 1.1).

Context 4

Operational police officers have always have been geographically mobile and policing is an interesting example of enterprise mobility, being one of the most advanced areas for the deployment of mobile services. The two-way radio system was first used outside the military by the Chicago police force during the 1930s prohibition era of emerging organised crime (Agar, 2003). Since then, police forces across the world have embraced mobile voice and data services as means of collecting intelligence, distributing information to officers in the field and co-ordinating efforts. A large body of research has, from a range of perspectives, studied policing (for example, Bittner and Bish, 1975; Manwaring-White, 1983; Klockars, 1985; Ericson and Haggerty, 1997; Manning, 2003). Several research efforts have also explicitly studied the role of information technology in policing, such as: the everyday work of police officers supported by a range of information technologies (Manning, 2008); the role of information technology as a means of holding the police accountable (Ackroyd, 1992); the role of intelligence-gathering technologies (Manning, 2008); mobile email in policing (Straus *et al.*, 2010); mobile technology and police practices in policing (Sawyer and Tapia, 2006); and the need for the co-evolution of technological and organisational interventions in order for police forces to benefit from technological investments (Garicano and Heaton, 2010).

Mobile police officers serve the role of symbolic representations of state power, as expressed well in a passage from the *Quarterly Review*:

> The baton may be a very ineffective weapon of offence, but it is backed by the combined power of the Crown, the Government and the Constituencies. Armed with it alone, the constable will usually be found ready, in obedience to orders, to face any mob, or brave any danger. The mob quails before the simple baton of the police officer, and flies before it, well knowing the moral as well as physical force of the Nation whose will, as embodied in law, it

represents. And take any man from that mob, place a baton in his hand and a blue coat on his back, put him forward as the representative of the law, and he too will be found equally ready to face the mob from which he was taken, and exhibit the same steadfastness and courage in defence of constituted order. (1870, pp. 90–1)

In a similar manner, the visible ecosystem of mobile devices and services applied by the police to co-ordinate their efforts also represent to the citizen the symbolic power to collect intelligence and to act in a co-ordinated manner upon it (Pica, 2006, Chapter 2.1).

One of the key challenges for operational police officers is to translate the uncertainties of the incident they are about to engage with into their own understanding and calculation of risk. The availability of reliable information is the main resource in this translation of uncertainty into risk. In this sense, one of the key characteristics of the police force in terms of organised mobile working is the orchestrated capability of maintaining centrally stored information that in turn is continuously collected, updated and acted upon in a highly decentralised manner. A key to understanding this capability of modern policing is the carefully cultivated balance between emergent and planned technology performances where the central repositories of information and associated functions for providing access to the information are organised in anticipation of decentralised emergent needs.

There are significant differences in the mandate and organisation of policing across countries. While countries such as Italy, Spain and the USA operate a proactive, paramilitary police force, UK policing is still largely based on the 1829 Metropolitan Police Act, which stipulated that officers should be unarmed, uniformed and reactive (Manning, 2003; Pica, 2006). Policing in the UK is carried out using a broad range of roles, such as criminal detectives, officers with primarily managerial and executive responsibilities, scene-of-crime officers collecting evidence, control room staff, traffic officers and response vehicle officers.

5.2   Mobile policing and technologies

Response vehicle officers, such as John and Mary, form part of the operational police force. They attend immediate response incidents such as domestic abuse, burglaries and public disorder. Response vehicle officers also provide support in keeping the peace, escorting prisoners, looking for wanted people, patrol crime hot-spots, appearing in court as witnesses, collecting information about reported crime, advising and calming down victims, and generally providing a sense of police presence. Domestic disturbances represent by far the largest proportion of incidents attended by response vehicle officers and are related to a range of social problems. The key to understanding this particular kind of policing is less one of crime-fighting than of peacekeeping through engaging in social work at high speed with flashing blue lights (Sørensen and Pica, 2005). Whilst out and about, the officers typically interact with the control room, other officers and supervisors, the ambulance service, crime management centres, victims, offenders and witnesses. Response vehicle officers have a significant degree of discretionary control over their engagement with mobile services, either when driving alone or as a team of two officers. However, their work and work environment – primarily the police vehicle and incidents – impose significant constraints on them. They are required to keep in touch with current developments by listening to the police radio, which offers a constant stream of information in the background. Officers have a trained ear for constantly monitoring this stream of information in order to assess when particularly important information is conveyed.

Mobile services generally find their own place in the mobile ecosystem within the limited space of the police vehicle or on the officers' person. Parts of the mobility technology for policing reside on or near the body of the officers, shoulder-mounted or in utility-belts, some of it may be installed in the police vehicle cockpit and yet other parts can be distributed within the environment, such as the telecommunications infrastructure. In the police constabulary from which the incident above is taken, response vehicle and traffic officers use a range of mobile services (Pica, 2006, p. 89):

- *The MDT:* this computer, with a small colour touch-screen, is situated at the bottom of the centre stack between the front seats. The terminal connects wirelessly to the control room, other response vehicles, the Police National Computer (PNC) and local police databases. It is equipped with an infrared keyboard for intensive data entry. Some vehicles are equipped with a stripped-down version of the MDT.
- *Car-based and personal radios:* police officers are constantly connected with the control room through car-based and personal radios. In most vehicles these are two separate systems, but some cars allow the car-based radio to be routed to the personal radio when the officer leaves the vehicle.
- *Car-based and personal mobile phones:* each response vehicle is equipped with a standard mobile phone. Police officers' private mobile phones are also frequently used to reach officers during emergency calls, because it can be difficult to anticipate in which car officers are located. However, this raises issues of accountability, as these calls are not logged in the official system.

The work of response vehicle police officers can be characterised in terms of five types of activity representing the main elements in their rhythms of collaboration. The officers use different combinations of mobile service performances in each of these five types of activity, reflecting the different needs and urgency in co-ordinating with other officers, with the control room or in the officers' engagement with the incident (Sørensen and Pica, 2005; Pica, 2006):

1) 15 per cent of the time is generally spent standing by in the car before an incident, where officers focus on the active queue of incidents and on looking for wanted people. The MDT is the primary means of managing information and thereby collaborating with other officers, followed by the radio and mobile phone;

2) 25 per cent of the time is spent driving to an incident, gathering information about the destination, potential risks and the vicinity of other vehicles in case support is needed. Officers here both communicate with the control room over the radio and receive data streamed to the MDT;

3) 34 per cent of the time is spent taking action at the incident, as well as continuously gathering information about the unfolding situation and emerging risks, and establishing a positive identification of the participants in the incident. Here the personal radio is the primary means of information management, perhaps supplemented by the occasional use of the mobile phone to contact people outside the direct reach of the control room, such as witnesses or citizens who have reported the incident – in terms of policing, these bystanders (Ferneley and Light, 2008) can form an integral part of the collaborative arrangement;

4) 13 per cent of the time is spent driving from the incident, where again attention is drawn to the active queue. If someone has been arrested at the incident, driving from the incident will take the police vehicles to custody so that the arrest can be processed and the arrested individual incarcerated. The radio is here a primary means of information management for the officers when leaving an incident; and

5) 13 per cent of the time is spent standing by in the car after the incident, and the informational requirements here are also the active queue of incidents and custody status. Immediately after an incident, when the response vehicle has yet again returned to awaiting new incidents, the radio is still the primary means of interaction, followed by the mobile phone and the MDT.

Throughout their working days, John and Mary apply a diverse set of services across a diverse range of contingencies comprised of both emergent and stipulated technology practices. Their work is characterised by significant uncertainty, which calls for emerging decision-making. However, they are also subjected to public scrutiny to ensure accountability, which leads to a range of stipulated performances, for example, documenting searches of citizens in stop and search forms and recording incidents in police reports. A range of stipulated practices also governs communication over two-way radios. This ensures that the radio waves are not occupied with unnecessary interaction. Communication with the control room about incident information through the MDT is also mainly governed by stipulated technology performances as the communication is structured and links directly into the police incident database. Complex rhythms of technology performances serve the purpose of discussing and resolving mutual interdependencies in collaboration between the control room, operational police officers, senior officers, incident witnesses and others.

5.3   Constant coupling and rhythms of collaboration

When organisational members engage in co-located collaborative arrangements with a low degree of collaborative complexity, the negotiation of mutual interdependencies can to a large extent be accomplished through ongoing processes of mutual adjustment and direct supervision (Mintzberg, 1983). When John and Mary rush to an incident, they engage in such a form of mutual adjustment and direct supervision when rapidly driving the police car through the traffic.

Local routines

John and Mary's routines have emerged as flexible rhythms of collaboration honed through working together on a daily basis and through adapting professional and organisational practices. They have, for example, adopted the practice of the officer occupying the passenger seat offering both navigational and informational support. This fosters tight interrelationships of their activities, with little distinction between the needs of the individual and the team in the hectic phases of driving to and engaging with an incident. Technology performances ensure that they both have an equally good understanding of the situation they are facing, and their collaboration resembles that of other cockpit-based teams working onboard aeroplanes or ships with close-knit local interaction and sophisticated technology performances (Hutchins, 1995).

The immediacy of the relationship between intention and action implies the initial notion of users having immediate access to engage with, separate from and re-engage with technology (Dourish, 2001, p. 139). The cab driver Ray discussed in Chapter 4 can easily pick up the phone and put it back in its holder again when he wishes to contact one of the members of his informal network. The CEO Hiro spends his days constantly shifting between communicating with colleagues and clients through mobile email.

The practicalities of work play a role in the feasibility of particular mobile services – for example, the physical circumstances of work (Kristoffersen and Ljungberg, 1999) or work demands for one's full attention (Sørensen and Pica, 2005). When cab drivers are en route, they must direct most of their attention towards driving, even if to some extent they do multi-task. They can, for example, update their location on the computer system, speak on the mobile phone using a hands-free kit or make brief notes while temporarily stopping at a junction.

When engaging *in* an incident, the police officers need to actively engage *with* the incident. As one officer states (Sørensen and Pica, 2005, p. 145): 'When faced with a person, who potentially can hurt you badly, you want to look that person in the eyes and not stand there and stare into a screen.' The conflicting battle for their attention when switching between modalities, primarily between voice- and text-based interaction, makes it difficult to maintain both (Sawhney and Schmandt, 2000). This implies that the shoulder-mounted personal radio becomes the primary means of interaction. The persistent radio connection with the control room is generally the only remaining service in use during critical incidents. It is the only one of the services in their portfolio offering the optimal sense of balancing the need for remaining in constant touch with the control room whilst at the same time devoting their full attention to the unfolding incident.

The role of mobile services shifts when the incident is resolved and the officers either return to the office to complete paperwork, go to the arrest or wait around for further incidents. As police work can be characterised in terms of changing rhythms of time- and safety-criticality, the officers' use of mobile services to co-ordinate activities critically needs to align with these rhythms.

Negotiating local and remote collaboration

The mutual interdependencies in operational policing reach far beyond the confines of the police vehicle and require the officers to engage with other parties beyond the vehicle. The officers continuously engage with a variety of situations requiring the contextual balancing of conflicting demands from the local and the remote.

When driving to an incident John and Mary rely on remote interaction with the control room mediated by the two-way radio system and the MDT to which incident data is streamed. They also remain in constant contact with the control room during incidents in order to update their position and significant developments, rapidly gather information, co-ordinate efforts with other officers and request further assistance in case it is needed. The two-way radio system represents an unprioritised constant two-way connection between the police officer and the control room. The service offers fluid collaboration. However, the ongoing cultivation of collaborative barriers is necessary for this fluidity to work. As the airwaves are shared, strict discipline is applied to the use of the system, for example, through brief compressed messages. The officers also apply highly selective filtering of the stream of radio messages continuously broadcast from the control room and thereby maintain peripheral awareness of ongoing and upcoming incidents. Such practices are also documented in studies of traders and underground train control room operators (Schmidt, 1993; Heath and Luff, 2000).

Mobile push and pull collaboration

Traditionally, police officers would rely entirely on the control room pushing information to all officers via the radio. This fluid push established a shared peripheral awareness with all officers as to the current state of affairs irrespective of boundaries. The alternative represents officers in separate vehicles directly pulling information from central repositories. The advantage of a highly targeted and decentralised pulling of information, which makes sense to the local officers, is the direct availability of more comprehensive and tailored information. However, pulling tailored information reduces the overall awareness of the general state of affairs shared by all officers. This can then make ad hoc co-ordination difficult to achieve. In practice, the particular police force John and Mary work for will continuously seek to cultivate the balancing of fluid push with the interactional barriers represented by the localised pulling of information.

The particular shared practices of how this balance is achieved are informed by broader concerns than the immediate choices of individuals and groups. Comparing the particular police constabulary from the vignette above with one of its neighbouring constabularies reveals an interesting difference in their use of mobile services to achieve collaborative effectiveness. The neighbouring constabulary has a booming tourist industry, of which a large proportion is young people. As a result, the police force here has a larger 'buffer' of officers to deal with problems relating to the seasonal influx of partying tourists. This constabulary can therefore rely much more on traditional modes of policing where communication with the control room is the main mechanism for co-ordinating policing efforts. In contrast, the constabulary from which the case above is drawn does not have the same need to deal with seasonal variations in terms of demand for operational policing. The relatively lower staffing levels are in this constabulary being countered by a greater diversity of mobile services in order to render teams more effective by, for example: enabling closer-knit co-ordination of resources; making it easier for officers to co-ordinate in a decentralised manner; providing each vehicle with direct access to centralised databases to check on individuals and vehicles; supporting a higher degree of mutual awareness in vehicles of what incidents are emerging; and by directly pushing data to the in-vehicle MDT instead of only relying on vehicle interaction with the control room over the radio. This illustrates the need for careful co-evolution of the portfolio of services with the organisation of police work in order to obtain increased collaborative effectiveness. It also illustrates that the careful balancing of push and pull information management as a means of better utilising staff.

The police officers evoke performances that at times engage constant coupling with the control room and that at other times selectively engage rhythms of collaboration with shifting intensities of the negotiation of mutual interdependencies. Considering the organisation as a whole, understanding police effectiveness can reveal interesting insights into the effectiveness of technological interventions. It can, for example, further highlight the well-known finding within IS of the importance of considering technological and organisational changes in an integral manner (Leavitt, 1964). Such observation will also have continued relevance within the study of the role of information technology for policing, for example, illustrated by the factor-based analysis by Garicano and Heaton (2010), even if this type of study engages in highly abstracted analysis, only considering aggregate measures in which the effects of IT investment are measured through variables representing resulting organisational changes. In the case of enterprise mobility, the concern is to provide a mobile services portfolio, which can represent sufficient diversity to suit the ongoing cultivation of fluidity and barriers. It is problematic to measure productivity entirely through translating services into actual performances. It is less about whether or not mobile services are followed by changing practices and more about what these changes are and how they relate to arrangements rendering policing effective.

5.4   Case 5: Simon – security guard

The following two cases (5 and 6) represent different work domains using very similar technological arrangements to facilitate local-remote collaboration between mobile workers and stationary dispatch offices.

Vignette 5

Late at night in an industrial estate on the outskirts of Manchester, Simon, a security guard, is doing his nightly round at an electronics wholesaler's warehouse. Traditionally, Simon would upon arrival at the office pick up a plan outlining his inspection route and, during his 12-hour shift, patrol the client facilities according to the plan by registering control points using an a hand-held reader, also called a *torch*. This reader would synchronise data collected with a centralised system on a weekly basis. Simon would spend a considerable amount of time on the mobile phone to the supervisor when events delayed him. The supervisor would, in turn, spend 90 per cent of his time on the phone to security guards in order to discuss the current situation with clients. Simon is currently engaged in a real-life experiment aimed at changing this situation into one with increased interactivity and awareness between clients and the security firm through the application of RFID technology. Five managers and 23 users use 12 mobile phones with integrated RFID readers. Simon no longer needs to pick up the plan of his shift. He is sent a message on the phone directing him to the next control point where he waves his mobile phone over a RFID tag mounted on the wall and a message is automatically sent to a central server to update the change in location (study 5 in Table 1.1).

Context 5

This is an example of field force automation where the technology improves existing working arrangements and renders co-ordination more effective (Scornavacca and Barnes, 2008). The real-time updating of the guard's location supports collaborative and management practices operating at a finer level of granularity than was the case previously. As Simon's work was already characterised by a low degree of individual discretion, the RFID-based services are not seen as radically changing the conditions of work. Indeed, the improved remote awareness of Simon's position means that the co-ordination with the dispatch office is conducted much more smoothly. There is also much less need for managers to engage in telephone conversations with clients as an extranet offers clients access to real-time information of the inspection status of their premises. The managers can also generate status reports in minutes. Francis, one of the other security guards, expressed early worries about the system: 'Now [superiors] are able to watch every step I take' (Kietzmann, 2007, p. 132). These concerns are, however, relatively quickly replaced by a positive attitude as the need for constantly reporting one's whereabouts and the manual activities of updating the paper-based documentation of the inspection rounds are both eliminated. For Simon, the convenience greatly outweighs the drawbacks of more direct supervision and the system also directly manifests to the managers the extent of his work. This view is also expressed by Walsh, another security guard: 'Before I always had to justify and explain everything because I work on my own, now they [superiors] can see that I am doing good work and I have to spend less time on tedious tasks' (Kietzmann, 2007, p. 132).

The following example describes the application of a RFID-based system supporting the management of industrial waste containers.

5.5   Case 6: Winters – industrial waste lorry driver

Vignette 6

Grizzly Waste offers a broad range of industrial waste services, such as container-based waste disposal and landfill waste management. The latter involves highly specialised complex processes of gas extraction and leachate (liquid drained from landfills) reduction. In contrast, the former seems much simpler. However, the management of waste between client sites and landfills is quite complex. Winters is a member of a team of bulk carrier vehicle drivers operating out of Manchester. He services several commercial and industrial sites with both established clients that keep containers on-site, such as large bakeries, and temporary clients with a specific need, such as managers of construction projects. Full containers are tipped either directly into landfills or into transfer-stations, where loading shovels separate waste before it is brought to landfills or recycling stations by other trucks. Each container costs around £5,000 and the company has traditionally not had systematic knowledge of the whereabouts of its containers and has therefore not been able to centrally optimise their best utilisation.

As a result of a growing customer base, many dormant containers at transfer stations and budgetary constraints on investment in new containers, there were a decreasing number of containers available to drivers. Winters had traditionally enjoyed a high degree of discretion in the way he organised his work. The team managed the flow of empty and full industrial waste containers between client organisation sites and dumps, as well as maintaining the vehicles. Traditionally only a few aspects of work had been documented centrally. The availability of drivers was managed by a mobile phone system where the key codes were entered into a mobile phone permanently mounted in the truck dashboard when a container was collected, tipped or delivered. The team kept paper records for each transfer of waste collection point, the client's name, cargo weight and waste disposal point. Containers were not systematically managed centrally and they were merely identified by drivers by the company logo. Paper records did not include information relating to individual container locations, their history with different drivers, customers or sites. The drivers were therefore the only ones in the organisation with a full insight into the exact whereabouts of containers. As such, the dispatcher would need to call the drivers to establish their location in order to find the one closest to a particular job. A range of coping practices had emerged as a means of optimising the overall performance of the team. These practices included temporarily storing empty containers halfway between customers to save on transport time, borrowing containers from or lending to colleagues from competing firms, or producing elaborate excuses to avoid being called to faraway emergency jobs.

In order to improve this situation, Winter's team is involved in a large-scale real-life experiment tagging 135 containers with RFID tags and equipping five drivers with RFID reader-enabled mobile phones and tags to be used for signing in and out of work. When reading an RFID tag, the mobile phone automatically relays the tag information back to a central system at the company headquarters (study 6 in Table 1.1).

Context 6

As a result of this additional information about both containers and drivers, the dispatcher knows where the drivers are and if they are able to take on emerging customer requests. Tracking containers also maximise the availability of containers through better logistics and thereby minimise disruptions. This can also lead to a decreased need for investment in new containers. The most experienced drivers are of the opinion that tracking containers may be beneficial for the organisation but not for them, as it can result in an additional workload, and generally that the new system does not offer significant benefits. Younger drivers do not have the same concerns (Kietzmann, 2007).

Here localised fluid collaboration amongst drivers is replaced with centralised intervention based on stipulated technology performances. Some workers find this particular interactive type of mediated co-ordination problematic as they previously had considerable discretion in the organisation of work and will now have to collaborate with the central dispatch office through discussions of how to accomplish the work. They feel that the dispatch office has an over-inflated impression of its own abilities to gain an overview of the situation through the updated information fed by the RFID-based co-ordination. The embedded co-ordination mechanism automates aspects of the work previously conducted by the user and also provides the possibility of real-time interactivity, as opposed to data being recorded in the centralised system with some delay.

5.6   From batch-time to real-time reporting with RFID

The RFID-based system stipulates performances forging an automatic ongoing relationship between the continuous recordings of, for example, Simon's route transmitted to a central database. This decreases individual discretion as information about the position of Simon, Winters or the container is automatically revealed.

Security work

The advantage of the reduction in paperwork is highlighted as important, as the information can now be provided automatically (Kietzmann, 2007, p. 196), and the constant connectivity implies less effort being spent on frequent explanations on the mobile phone to the central control room (p. 132). In Simon's case, this implies a shift from the negotiation of mutual interdependencies through interaction symmetry between the security guard and the control room towards highly formalised and prioritised interaction between his handset and the central server. This shift removes the need for continual co-ordination between the guard and the control room concerning the state of the common field of work, and at the same time provides much more direct and granular stipulation of Simon's work.

Simon's RFID-enabled mobile phone essentially consists of a protocol triggering the display of the next location Simon must patrol once he has scanned the RFID tag at his current location. The co-ordination mechanism contains a classification of locations, which is mapped against the simple protocol of sending Simon to the next embedded RFID tag. As his work can be characterised by a low degree of individual discretion, the normal situation is one where a relatively narrow portfolio of specific mobile services will be expected to translate into a narrow series of intended performances. It is not expected that he will refuse to scan tags or will decide to alter his route according to his own preferences. The co-ordination mechanism is in this case close to a script determining his actions and thereby also the negotiation of mutual interdependencies between Simon and others inside and beyond the security firm. In case of an emergency, planned technology performances can be replaced by emergent performances dictated by the specific situation. However, even in such emergencies there will be attempts to counter-balance emergent actions and emergent technology performances through the establishment of procedures for planned interventions and associated planned technology performances. For example, Simon will be required not to engage with any suspected burglars or indeed to attempt to put out a fire. Instead, he will be required to immediately call for support and will seek to move away from any personal danger.

Managing waste

Even though much of the interaction is automated through a combination of a central system and distributed RFID-enabled mobile phones, the direct interactivity between mobile work and centralised planning creates increased real-time demands for truck drivers to document work. Winters argues that he used to do all the paperwork at the same time or that he would do some of it when waiting at a tipping station, at railway crossings or during coffee breaks. The new system requires him to complete 'paperwork' (on the mobile phone) as it emerges – even when there is not sufficient time. As a result, he will complete some of this work whilst driving (Kietzmann, 2007, p. 187).

The sudden transparency between the remote trucks and the central dispatch office implies that local practices are suddenly questioned. As Schaitel, one of the other industrial waste lorry drivers, puts it:

> Through the RFID, they [the managers] can tell when I dropped off a container and when I picked up the next. We always pinch 20 minutes here and there; we drive twelve hours per day and need breaks. In the meanwhile, the truck is empty – they don't like that. It's always been like that and they have always turned a blind eye if we did not overdo it. But now they have proof and they have to act on it. They ask us to send an SMS when we go on break so that they know that we're not lost or in an accident. We never did this before and we need to figure out a way around this. (Kietzmann, 2007, p. 188)

The interactivity of the new system alters the collaborative dynamics between mobile workers and the central dispatch offices. In the case of Simon, this is seen as positive. For Winters, it is perceived as less positive. Turner (1984) documents a similar change in user dynamics when an online system for managing social security claims replaced a batch system. The resulting immediate response altered not only the working conditions for the users but also the expectations of the claimants of instant decisions.

RFID relationships

One of the primary changes with the RFID system is replacing memoryless encounters with an ongoing managed relationship. The RFID readings are automatically fed into the central database, which then becomes the constantly updated externalised memory of work in progress. The mediating artifact and associated protocols thereby reduce the collaborative complexity (Carstensen and Sørensen, 1996). The RFID system creates awareness of selective aspects of the entire work process (Zammuto *et al.*, 2007) – the movement of security guards or waste containers to participants beyond the team. The resulting collaborative transparency allows remote managers not to only question previously hidden mobile working practices but also to assume that this particular aspect that is rendered transparent is all that is needed for effective collaboration.

This is the cause of conflict as local, individual and team discretion to organise work is challenged. The use of mobile services can support the transparency of activities beyond the immediate co-present group in similar ways to ordinary groupware technology – for example, a design team suddenly realising that a remote group of top management is following its progress (Ciborra, 1996). The resolution in this latter case was to establish collaborative boundaries allowing only team members to gain detailed insights into team discussions. Here the move from collaborative fluidity was augmented with barriers shielding the core team. In the waste management example, collaboration within a clearly bounded collaborative arrangement was rendered fluid and open through the use of mobile services. The central system models the interactive behaviour of one aspect of the common field of work – the position of the container.

5.7   Symmetry and asymmetry in collaboration

Embedding prioritisation of the interaction by filters and notifications transforms the assumption of interaction symmetry of a simple standardised connection into one of asymmetry. The changes associated with replacing direct telephone connections and batch processing with instant reporting and stipulation of remote behaviour are examples of a prioritised service implementing interaction asymmetry between the local and the central as the service no longer assumes the equal status of each party. Technology symmetry leaves the cultivation of barriers in the form of interaction asymmetry practices to the user. Technology asymmetry directly supports the user in managing interaction prioritisation.

Interaction asymmetry can support collaboration through interaction filtering and the awareness of incoming interaction (Ljungberg, 1999; Ljungberg and Sørensen, 2000). This provides support to individual mobile workers in the orchestration of their interaction with others. The CEO Hiro (discussed in Chapter 4) defines interaction asymmetry through rules prioritising incoming requests for interaction and attaches different actions to these along with the organisational practices converging interaction into mobile email.

The translation of technology symmetry and asymmetry into technology performances is far from simple and straightforward. Technology performances occur in social contexts and here both shape and are shaped by the richness of these contexts. The promise of technology symmetry between two mobile phones through a standardised unfiltered voice connection will inevitably be translated into a social context of interaction asymmetry. The recipient can refuse the request and not pick up. Even if the call is taken, the performance will be subjected to inherent asymmetry of the social adaptation (Goffman, 1959).

Co-ordination mechanisms contain protocols stipulating and prioritising interaction and can therefore directly support prioritisation in the resolution of mutual interdependencies through modelling these using technology asymmetry (Schmidt and Simone, 1996). Simon's co-ordination mechanism prioritises the interaction and reduces the degree of local control necessary for him to coordinate his work. The co-ordination is largely taken care of by the stream of SMS messages between Simon's mobile phone and the central server.

A mobile phone user's customary glance at his or her phone when it rings in order to ascertain the identity of the caller is a well-known example of respondent awareness of a request for interaction (Ljungberg, 1999; Ljungberg and Sørensen, 2000). However, it is also conceivable that other aspects of the desired interaction than the simple request could be displayed on the screen – for example, subject and urgency (Ljungberg and Sørensen, 2000; Wiberg and Whittaker, 2005). As demonstrated by Wiberg and Whittaker (2005), a co-ordination mechanism can take over part of the *outeraction* (Nardi and Whittaker, 2000) of negotiating when interaction should take place, given some information relating to individual preferences and schedules. However, as also noted by Wiberg and Whittaker (2005), this may prove difficult to accept as experiments show a reluctance to delegate the responsibility to technology since people felt a social obligation to return calls. This is entirely in line with Goffman's (1959) assertion of the social character of communicative acts as a reflection of a person's identity.

A group of highly interdependent workers engaged in intense unprioritised interaction will most likely wish to explicitly manage group membership. This defines collaborative barriers of interaction asymmetry within which participants may engage in fluid and focused collaboration. Instant messaging buddy lists represent an example of such a technological stipulation of priorities. Cultivating interaction boundaries support the group in session management, allowing desired participants to enter and keeping undesired participants from entering. What for the outsider of a group may impose an asymmetric constraint is for the insider an affordance.

For John and Mary, mobile services provide a variety of collaborative support mechanisms – for example, the timely push of dedicated information through asymmetric channels to the in-car MDT; flexible symmetric interaction encounters through the mobile phone; and, critically, the support of a continuous individual relationship with the control room through the personal radio. The collaborative complexity of response vehicle police work is greater than for security guards and waste lorry drivers, in that it is time- and safety-critical, the work deals with human beings in extreme situations and it requires the careful negotiation by individuals and teams of the boundaries between adhering to formal procedures and guidelines and, on the other hand, exercising discretion.

The affordances offered by mobile services become integrated in everyday organisational life and the outcome of their use will be the result of complex socio-technical processes characterised by human and material agency (Leonardi, 2011). The social shaping of expectations and performances, for example, results in only a small likelihood of being contacted on the mobile phone during weekends; an organisational guideline of people answering emails within a couple of hours; or expectations of everyone being signed into the corporate instant messaging system throughout the working day. Whatever the practices, social rituals or organisational rules, each individual will be left to grow his or her own personal relationship with his or her particular portfolio of services and to apply a variety of coping strategies to manage his or her particular collaborative practices. Khalid, the mobile trader, can out of hours easily negotiate exposure limits or engage in other forms of co-ordination with other traders through mobile phone calls. However, he tries to avoid such contact unless it is deemed absolutely necessary in order to avoid disturbing colleagues who are also both on and off work at the same time. Hiro applies such practices when seeking to persuade others to use email when contacting him – as the founder and CEO of the company, he clearly has significant power and clout to arrange the world around him in a desirable fashion.

5.8   Case 7: Jun – town planner

Vignette 7

Jun is 38 years old and since 2000 has had his own independent town planning practice that mainly works with small- and medium-sized Japanese municipalities. These are located around Japan in rural areas hundreds of miles away from large cities. Jun engages in projects with a range of people, such as developers and other consultants. However, he is the sole employee in his practice. Town planning projects typically require that project members visit the specific site in question for development. For Jun, visiting the site and seeing it with his own eyes offers invaluable insights and is therefore deemed to be crucial for his business. He sees his flexibility in locating himself where the project is situated as an advantage that he can offer his clients. This means that he relies critically on mobile services as a means of setting up his office wherever a project takes him. His technological set-up consists of a large notebook computer situated in his Tokyo office and a compact sub-notebook that he brings with him when he works away from the office. He typically engages in two different patterns of movement through his work. He will spend a considerable amount of time on-site with clients situated far away from his office in Tokyo. This involves frequent long-distance travelling by train or plane. When based on-site for a period, he will engage in extensive local commuting by taxi, bus or walking. This kind of local travelling is also frequently the case when Jun is back in Tokyo, as here he will often need to meet other project members or visit clients around the city (study 7 in Table 1.1).

Context 7

Japanese professionals have traditionally been employed in large organisations (Aoki and Dore, 1996). However, an emerging itinerant workforce of professionals are organised in small companies or work on their own (Perkin, 2002). Organisations are increasingly attempting to manage risks and establish employment flexibility by relying on itinerant workers of various kinds (Malone and Laubacher, 1998; Barley and Kunda, 2004).

Itinerant work can consist of both stationary and mobile working, and relies significantly on mobile services as a means of engaging in collaboration. The study of 63 Japanese professionals reveals intense notebook computer and mobile phone use (Kakihara, 2003). In Jun's case, mobile services are instrumental for him to work independently as a single person engaged in a project working with other parties, without employing administrative support staff.

For cultural reasons it is not seen as acceptable for individuals to unsolicitedly engage in the promotion of their professional services. If an individual contractor needs work, the work will largely have to come through requests from others. This is quite contrary to other cultures, where it is seen as quite acceptable to openly offer one's services (Barley and Kunda, 2004). However, both cultures lead to a significant amount of time being spent socialising and networking to secure future work (Kakihara and Sørensen, 2002; Nardi *et al.*, 2002). This implies that work for small organisations and for individual professionals is primarily found through social relations and as a direct result of past projects. Such practices have also been confirmed for independent Greek software engineering professionals (Voutsina, 2008).

5.9   Individual and collective working

The following text discusses the distinction between enterprise mobility as a means of optimising collaborative arrangements as opposed to significantly transforming these.

Optimising collaboration

All the cases of enterprise mobility discussed up to the case of Jun have essentially illustrated changes to existing roles and practices through mobile technology performances. Hiro, the roaming executive, may conduct his work whilst out and about in Tokyo, but he essentially still conducts executive work with similar arrangements beyond his mobile habit. Ray's work is not essentially different from how taxi work has been conducted for decades in London. However, he is able to connect with both colleagues and family members while driving and can also receive jobs on his computer-cab terminal. For Khalid, mobile trading out of hours is highly individualised and obviously different from the collective work of trading during the day. However, he is still very much part of the same collaborative arrangement day and night, even though the negotiation of mutual interdependencies mostly occur during the day. John and Mary conduct their policing work differently from traditional policing by deploying a range of specific services. However, in terms of the organisational arrangements of roles and responsibilities, there are distinct similarities to traditional forms of policing. For Simon, it becomes easier to co-ordinate his progress with the central control room. For Winters, collaboration with the dispatch office changes significantly as previously localised practices are suddenly challenged. Common to all of the cases discussed above is the fact that the transformative potential of mobile services is deployed within organisations without the fundamental organisational arrangements being challenged.

Collaborative transformation

The case of Jun, the town planner, is different as the entire organisation of work is based on the removal of immediate collaboration within a professional services firm, which thereby removes a host of mutual interdependencies. Instead, the interdependencies essential to re-development projects are nurtured intensively by Jun physically residing where a current critical project is located. Jun perceives that being an itinerant project worker provides him with a level of agility he would not have in the context of a larger organisation.

Jun used to work for a small town planner consultancy firm and found working with local councils around the country exciting. However, he argues that this led him to conduct a range of time-consuming administrative activities unrelated to his main interest of working at local sites, and that being independent minimises this administrative overhead (Kakihara, 2003, pp. 139–40). He explains that working alone means the ability to freely move wherever he finds it necessary, as the traditional restrictions imposed on employees in large firms are removed:

> In a firm, you need to have a confirmation from your boss for the trip. Then you need to apply for a travel budget and wait for a reply from the accounts section. Once back to your office from the trip, you have to make a report of the trip, which nobody reads. All of these cost immense time. But it is quite normal in this business that your client in a rural area hundreds of miles away wants you to come to their place the next morning. Also it is normal that during a trip you will need to visit other places unexpectedly due to the

client's need. You can't act flexibly in such situations as long as you work in the firm, ending up with losing business opportunities. Working independently now, I can move flexibly and my clients know it. Mobility, I would say, is the most immediate and powerful advantage of working independently. (Kakihara, 2003, p. 142)

Jun also actively engages in resolving the dilemma of short-term business concerns and his passion for making a lasting impact through longer term concerns for the needs of local areas for town planning. He argues that being a good town planner requires being stationed for a considerable amount of time on-site and that projects in reality take years or decades, not the six months normally allocated to them. This was not viable when he worked as an employee in a large firm; 'So it was a big dilemma for me as an employed town planner, between being efficient as a member of the firm and being effective for the local people. Then, as you see, I chose the latter' (Kakihara, 2003, pp. 140–1). In terms of the overall benefits of his arrangement with the client organisations, he states that: 'independent consultants like me can participate in the project for quite a long time and therefore can create intimate relationships with the local government [...] The point is, we are adaptable, the big firms are not' (p. 143).

Individualisation of collaborative efforts

However, the individualisation of work, such as in the case of Jun, can create additional overheads for the individual in his or her interaction with other parties. Whilst Jun may have removed the need to work in an organisational setting, he still has a need to engage in collaboration with others in urban development projects and to discuss future business opportunities with colleagues and potential clients. He also needs to fulfil tasks that the division of labour would have assigned to others.

Jun explains his use of his mobile phone for this purpose as a means of flexibly managing these demands. He forwards all calls from his office to his mobile phone, and calls are automatically forwarded to voicemail if he does not pick up. However, clients also tend to call his mobile directly, and he continues: 'When we didn't have the mobile phone, workers like me would have had to either hire a secretary to receive phone calls or just give them up completely when going outside for a long time. Now, the mobile phone solves it' (Kakihara, 2003, p. 147).

The duality of technology performances and mobile working enables existing collaborative arrangements to become more effective through changes to the technology performances. It also implies a more explicit ordering of collaborative arrangements so that these are directly amenable for support through the use of specific technology performances. The traditional organisation of complex mutual interdependencies in projects is an example of the organisation of activities, so these are amenable to the management through a variety of technologies in order to map progress and determine collaborative constraints. However, even at a lower level, within projects, the re-design of interdependencies and the modularisation of tasks make it possible to individualise efforts that previously would rely on intense collaborative efforts. This has been demonstrated in the case of Greek software developers working on their own or in small companies where the delineation of work tasks for the individual was supported by a modular architecture of the common field of work. This allowed an increased individualisation of the effort subsequently assembled into a whole (Voutsina *et al.*, 2007; Voutsina, 2008). In this sense, one of the effects of enterprise mobility is an increased opportunity and pressure to work alone, thus matching the social life of 'bowling alone' (Putnam, 2000).

Jun arranges his mutual interdependencies so that these can be flexibly negotiated through his physical presence. The project is a primary mechanism for the organisation of these interdependencies. There is in this example a conflation between the institutional arrangement of an individually owned company and the individual role in the projects in which he participates. While this arrangement is similar to that of Ray, the taxi driver, it is not primarily mobile services that afford Ray his way of working but rather the nature of his work and 'the Knowledge'. For Hiro, the CEO, advanced use of mobile services allows him extended flexibility in where he chooses to be, but the complexity of his mutual interdependencies with others is not radically altered, merely the means by which he engages with these. For Jun, mobile services afford a high degree of individualisation of his efforts so the mutual interdependencies can either be negotiated when he is physically at the town hosting the project or reduced to a level that is possible to negotiate through mobile services.

In terms of the institutional arrangements, Jun and Ray are the only two workers discussed who work not only on their own but also for themselves. All other cases in this book are mobile workers who work as part of an organisation with the overhead of mutual interdependencies in the collaboration this entails. However, what is characteristic for almost all of the individuals studied is that a significant proportion of the work is conducted in solitude: Hiro seeks inspiration by wandering the streets of Tokyo; Simon and the Winters work on their own; and Khalid works on his own outside of normal trading hours.

While all the cases illustrate how solitary individuals conduct work, they also highlight the importance of work as a complex collaborative endeavour constructing, negotiating and resolving interdependencies between participants (Schmidt, 1993). Mobile working can exemplify one of the paradoxes of virtual teamworking, in that interdependent teamworking is increasingly conducted through individual contributions (Dubé and Robey, 2009).

5.10   Transparency in collaboration

The co-ordination of work activities within teams implies the management of mutual interdependencies (Schmidt and Bannon, 1992; Malone and Crowston, 2001), which can be resolved through combinations of mutual adjustment, standardisation and direct supervision (Mintzberg, 1983). Mobile services can both support emerging and stipulated performances of mobile team members co-ordinating activities through: 1) lightweight support for mutual adjustments by affording direct access to synchronous and asynchronous connections; 2) the ability to push systems of standardisation to the individual mobile worker, allowing teams to operate based on a shared understanding of these standards; and 3) offering extensive means for mediated supervision. Through such support, mobile services offer both the promise of increased transparency in distributed collaborative efforts and opportunities for rendering collaborative arrangements more opaque as both individuals and groups may no longer be able to gain an overview.

In the two RFID cases (5 and 6) the technology performances resulted in greater transparency, whereas engaging in mobile foreign exchange trading in Khalid's case (case 3) led to a reduction in transparency, which in turn was countered by the requirement of documenting trades. From an analytical perspective, focusing on the co-ordination of mutual interdependencies emphasises the cultivation of horizontal co-ordination of work amongst peers as opposed to the vertical aspects of managing work through command-and-control decision processes (Schmidt and Bannon, 1992; Malone, 2004). However, such a distinction is analytical as the organisation of individual teams can mix horizontal and vertical aspects depending on the extent to which a team relies on a project leader (Perlow *et al.*, 2004).

Police forces have in general embraced a diversity of mobile voice and data services as a means of co-ordinating distributed activities, intelligence gathering and distributing information to officers in the field (Manning, 2008). Maintaining an overview of the collaborative arrangement is deemed essential for both the highly distributed operational police officers and for control room staff and senior police officers. This overview not only ensures the ability to rapidly deploy additional resources when needed but also ensures accountability of the actions taken. The MDT offers a rich source of appropriated information pulled by individual officers before engaging with an incident, although at the cost of providing other officers with a chance to obtain awareness of the situation. This illustrates the constant balancing between rendering relevant aspects of the collaboration transparent and others opaque. In the study of the lorry driver Winters in case 6, the central system directly modelling the interactive behaviour of the common field of work – containers – produced such transparency of teamworking as to allow remote dispatchers to not only question previously hidden mobile working practices but also to assume that all that is needed for effective teamworking is this particular transparency.

5.11   Cultivating collaboration

Enterprise mobility implies the assumption of geographical mobility, which implies a departure from traditional, fixed, co-located working practices. Traditional studies of team interaction assume that key members of the team are located within the same local area or building (Perlow *et al.*, 2004; Felstead *et al.*, 2005).

The traditional understanding of meetings in organisations assumes that these consist of a number of people gathered in a room in order to engage in a lengthy session, which has been planned in advance (Whittaker *et al.*, 1994). However, measured in terms of frequency, meetings are typically organised in terms of micro co-ordination (Ling, 2004), where two people engage in a brief unscheduled meeting continuing a discussion on the basis of previous discussions (Whittaker *et al.*, 1994). Collaborative activities are therefore typically complex patterns of interweaving multi-threaded brief interactions (Wiberg, 2001; González and Mark, 2004).

Amplifying interpersonal connections through a variety of networking services (Mathiassen and Sørensen, 2008), such as mobile email, voice calls and SMS messages, can support the management of an increasing intensity of such micro-co-ordination at the same time as these technologies can fuel further increases in interaction. In their study of the organisational adoption of Lotus Notes, Schultze and Vandenbosch (1998) demonstrate that information technology can both increase the ability of users to process information and the amount of information they are subjected to.

Organising collaboration implies the cultivation of mobile practices, i.e., organising where work is conducted, with whom and how mobile services support the working arrangements. Here the cases display significant variation between the highly independent taxi driver Ray, who exercises discretion as to when and where he works, and Simon the security guard, who works according to a centrally stipulated plan. In distributed collaboration, bringing team members together can foster socialisation and mutual trust, facilitating improved teamworking (Dubé and Robey, 2009) and engaging members in critical learning processes (Orr, 1996). Mobile information technology can support ongoing processes of socialisation and learning through a technologically mediated connected presence (Licoppe, 2004; Ling, 2008), for example, as in the case of London Black Cab drivers maintaining close-knit communities through frequent mobile phone conversations exchanging relevant information and socialising (Elaluf-Calderwood, 2008).

Much of the existing research in computer support for collaborative activities emphasises the symmetric and reciprocal nature of collaborative work, but does not discuss the asymmetric aspects of collaborative activities (cf. Schmidt and Simone, 1996; Heath and Luff, 2000). This emphasis on the mutuality of interdependencies can largely be attributed to the desire to explain the realities of unfolding work activities without too much attention being given to formal hierarchical arrangements that simply impose normative organisational views on how work ought to be understood and not how it actually is in practice.

Formal constructs, such as co-ordination mechanisms, serve the purpose of offering opportunities that stipulate the affordance of interaction asymmetry. As opposed to other affordances supporting distributed collaboration, such as support for maintaining mutual awareness and shared workspaces enabling the exchange of digital objects, the co-ordination mechanism cannot be designed without some assumptions of interaction asymmetry. The purpose is to reduce aspects of the complexity of negotiating mutual interdependencies through affordances prescribing central aspects of their resolution. This inevitably involves some form of prioritisation – either embedded within the mechanism as a technology affordance or provided as part of the associated principles for its operation.

To give an example, the co-ordination mechanism used by Khalid to document deals conducted after normal trading-floor hours requires him to leave an answer-machine message serving the purpose of documenting deals. The mechanism is not a symmetric two-way communication channel but a one-way recording channel documenting activities for subsequent use. Similarly, when the response vehicle officers John and Mary inspect the list of active incidents constantly updated on their mobile data terminal and select to engage with a particular incident, they do not have a range of possible options, but are merely required to select the incident. In that instance, it is allocated to them and is not available to others. The interaction sought is orchestrated through affordances standardising and prioritising interaction by reducing the individual's degree of local control.

5.12   Summary

This chapter has analysed the role of mobile services and performances in terms of constant connectivity and rhythms of collaboration, the role of interaction symmetry and asymmetry, and the shifts between individual and collective working arrangements. Mobile technology performances support the negotiation and resolution of mutual interdependencies in collaboration through shared information repositories, support in gaining awareness of the state of affairs, and co-ordination mechanisms scripting or supporting distributed activities. Collaborative technology performances can both render collaboration more transparent and more opaque. Six of the seven cases of enterprise mobility demonstrated both the evolution and optimisation of working practices with mobile information technology. One case study demonstrated the transformation of corporate employment into self-employment through the combination of mobile services and extensive geographical mobility. In all cases the cultivation of fluid collaboration and of collaborative barriers through technology performances is a key activity. The following chapter emphasises control and discusses the role of mobile technology performances for managing effective organisational interventions in mobile working.

6
Control – Effective Interventions

This chapter is concerned with technology performances related to the control, planning and general management of mobile work as performed by or subjected to mobile workers. Hughes *et al.* (2001, p. 63) argue that the increased virtualisation of teamworking 'creates managerial problems in the form of monitoring and control'. Most mobile work is not directly observable and therefore requires the use of a standardised symbol language modularising and mediating the interaction (Schmidt, 1993; Kallinikos, 1996). The aim is not to explore the complexity of effective organisational interventions in order to manage work in general, as this would require embracing the entirety of the organisation studies discourse. Rather, the aim is to identify salient issues of technology performances for effective intervention in mobile working, such as the tensions between remote control and local discretion, shifts in direct observation and indirect control, and the role of trust in mobile working. The issue of remote control and local discretion is explored in the case of a food delivery driver. The discussion of a health sector professional engaging in work-integrated learning exemplifies the issue of direct observation and indirect control. The chapter synthesises findings into the discussion of technology performances supporting the cultivation of decision flows and organisational boundaries.

6.1   Case 8: Jason – delivery driver

Vignette 8

Jason works as a delivery driver for Foods International and he delivers the range of products needed by small restaurants and fast-food outlets, such as frozen and fresh ingredients, paper plates, disposable cutlery, napkins, soft drinks, etc. Like many of his colleagues, he has only worked with the company for a relatively short period of time. However, the systems he relies on in his daily work are designed to guide him through his working day.

He receives updates to his mobile phone on the readiness of customers to collect their orders and updates on the traffic situation. Foods International seeks to ensure tight control over operations through the use of logistics technology, which stipulates Jason's delivery route. If he finds a customer's shop closed at the time of delivery, he uses the company's mobile phone to find out from Customer Services the whereabouts of the customer. The answer determines whether he will have to return at a later time or wait around a bit longer. The company rules explicitly state that Jason must use the company Customer Services department as an intermediary in case of problems delivering goods or when there are discrepancies between the order and the delivery. However, this does not always work out. For example, when a customer is not available to take the deliveries, Jason at times overrides the formal procedure and decides to either park near the shop and wait or to make an unscheduled return at a later point on his delivery round. He has also given his mobile phone number to a few of the customers he frequently serves. This enables the customers to contact him directly without having to go through Customer Services. As this communication occurs without being centrally recorded, the organisation is unaware of any such separate arrangements made between Jason and the customer. In the eventuality of Jason leaving the firm, the customer will then also be calling someone no longer working for Foods International or, if he or she is calling a company mobile phone, someone other than Jason. Jason's route is not represented in the central logistics system, for example, through GPS data. This provides him with the opportunity to engage in local adaptation to resolve emerging situations. The causes of emergent actions can be the traffic flow, unavailable customers or self-determined desire on the part of Jason. It is not possible for the organisation to unambiguously verify or falsify Jason's statement about his actual position or about the emergent situations that may have required a change to the delivery schedule (study 8 in Table 1.1).

Context 8

Foods International is a family-owned company established in 1988 as a wholesale supplier to small restaurants and fast-food outlets by the owners of a chain of restaurants. The company employs around 700 people in the central London depot and four other locations across the UK. It also seeks to deploy advanced logistics systems in order to ensure smooth operations. This technology plays a critical role in orchestrating operations across a 24-hour cycle beginning in the late afternoon when a fleet of lorries delivers goods to the warehouse. The orders taken up until 5:30 pm will almost immediately after this deadline be semi-automatically processed into planned delivery routes for the next day based on a range of criteria, such as customer postcode, lorry size, driver familiarity with the route, total volume and tonnage of delivery, value of orders and the general traffic situation (Boateng, 2010, pp. 128ff). The planned routes are used to calculate pick-lists for warehouse staff to pick orders from the shelves and pack the lorry in the most appropriate order compared with the delivery schedule. Orders are picked and lorries are packed throughout the night, ready for delivery by 6 am in the morning by the company fleet of nearly 200 delivery lorries.

6.2　Remote control and local discretion

Control can be characterised as the negotiation of mutual interdependencies related directly to collaboration itself – vertical monitoring and intervention of the horizontal collaboration (Schmidt and Simone, 1996, p. 159). This implies a vertical nesting of concerns where horizontal collaboration at one level becomes the common field of work at the next level.

Creating pockets of discretion

This vertical perspective on work is not necessarily directly related to organisational hierarchy, although it may often be so. Actors can also play significant roles in both the co-ordination and management of mobile working – the role conflation in the case of Ray the taxi driver implies that he both negotiates fares with customers and schedules when he is working.

In Jason's case, the horizontal co-ordination of deliveries between himself, customers and the control room is subjected to a range of formalised controls embedded into the technology or represented by organisational procedures. These control activities are defined and are sought to be enforced at an organisational level beyond Jason. Mutual interdependencies between delivery drivers and contact centre staff taking orders from customers and the warehouse staff assembling orders and packing lorries for delivery are largely co-ordinated by a centralised organisational system. This system is also programmed by the route planner to schedule delivery routes.

Jason is continuously exposed to the need for discretionary decisions to counter emerging situations that change the schedule. He critically relies on the availability of the managers to get the delivery signed off and is also frequently at the mercy of a highly variable traffic situation in and around London.

When emerging situations require discretionary decisions, the organisation seeks to place this discretion centrally within the Customer Services function. However, as the central system does not have full knowledge as to the emergent situations and indeed Jason's exact location, he is subjected to a continuous balancing act between central attempts at exercising remote control and his perceived need to maintain the right to exercise local discretion in order to manage the emerging situation. This operational freedom, which Jason carefully manages, results in technology performances balancing the closely stipulated principles for reporting to the central customer contact centre in order to centralise decisions with the need to smoothly deal with emerging customer situations, for example, restaurants not being open at the required delivery time.

Jason's emerging performances are an immediate way of addressing unplanned constraints. However, they can also represent the alignment of his interest in getting on with work while at the same time providing good customer service, for example, when he gives his mobile phone number to a restaurant manager. As in the industrial waste example in case 6, the localised emergent situation can be the subject of localised practices in order to resolve the situation, even if the outcome is not optimal for the organisation.

Local discretion and relationship control

The primary task of police officers John and Mary is to deal with emerging situations and therefore they need significant discretionary powers to do so. However, they are in much more interactive contact with the control room and with colleagues through a variety of communication technologies and are therefore also more easily subjected to direct remote control. But such direct control is not a dominant feature, as the response vehicle officers are not armed. For the armed response officers, direct remote control is much more proceduralised so that officers, for example, only open the weapons safe in the vehicle once they have been given an explicit order to do so by their superiors in order to ensure the proper allocation of responsibility (Pica, 2006, p. 143).

Response vehicle officers are required to fill in a separate report each time they decide to stop and search an individual, along with the normal reports and extensive paperwork required when someone is arrested. This documentation serves the dual purpose of holding officers to account for their individual actions as well as gathering information across constabularies.

For Hiro the CEO, Ray the taxicab driver and Jun the town planner (cases 1, 2 and 7), control plays a significantly different role in the operational unfolding of mobile working. Hiro is a CEO and company owner who is not directly subjected to control but rather exercises control over others if anything. This is done through constant interactivity and through the organisation of activities into projects with budgets, resources, goals and milestones. Similarly, for Jun, who works for himself, control is primarily exercised within projects. Projects are a common mechanism for control within and between organisations (Kunda, 1992).

Ray is the most operationally independent person of the mobile workers studied, as he owns the cab he drives. He is, however, subjected to institutional control from the Public Carriage Office (PCO) in order to maintain his taxi licence. The computer cab company Ray is associated with monitors and controls his performance in relation to completing his monthly quota of runs through the system. Finally, by acquiring 'the Knowledge' and practising the application of it, he has internalised a complex means of balancing planned and emerging actions.

6.3   Case 9: Yin – peri-operative specialist practitioner

Vignette 9

Yin used to be a nurse and she was very good at her job, so she decides to further specialise for even more challenging work. She wants to qualify as a peri-operative specialist practitioner (PSP) assisting surgeons in pre- and post-operative work. This involves on-the-job training for one year at the hospital where she works. This morning she is following surgeons doing their rounds. An essential part of her theoretical learning and practical training is done at one-week sessions every six weeks in London. Here the main co-ordinator of the programme is keen to follow and record Yin's progress when she is back at her hospital. This is essential for both providing feedback on the learning process and for documenting progress to ensure subsequent certification. She is therefore provided with a PDA with proprietary software to record her conduct and the outcome of each session back at her hospital. The PDA docks into a foldable keyboard, making it easier to enter text and connect wirelessly to the Internet. The PDA also comes with an address book application, a calendar application, email and scaled-down versions of Microsoft Excel and Word. Yin can use these to record information, to document clinical and learning activities, to note reflections immediately after a learning activity, to share information and to transfer relevant data to the learning centre in London. Yin is supposed to continuously populate the Actions Log Database detailing patient encounters on the wards by selecting clinical actions from a pre-defined list. She is also expected to fill in a Reflective Journal according to a set of template headings. However, she finds this very difficult to accomplish, as the PDA constantly seems to get in the way of learning and working. Nevertheless, the PDA comes in very handy for her own personal information management and she also frequently uses its built-in medical dictionaries (study 9 in Table 1.1).

Context 9

When the European Union introduced the European Union Working Time Directive (EUWTD) in 2003 many UK professionals exceeded the maximum of 58 working hours per week. In particular, junior doctors in the UK NHS were breaching the Directive with an average of 72 hours per week. The gradual introduction of the Directive, which was fully implemented in August 2004, meant that more junior doctors needed to be trained – a lengthy and expensive process. In order to alleviate this problem, the NHS Changing Workforce Programme at the Department of Health started a number of pilot projects. The PSP project is one such attempt and is aimed at introducing a new professional role in surgical teams providing pre- and post-operative care, with this new profession assuming some of the responsibilities previously held by junior doctors, in particular, peri-operative management for elective and emergency surgical care with diagnostic and procedural skills.

Yin is part of the first group of 12 PSP trainees recruited from existing medical staff such as nurses and operating theatre technicians, who are taught, amongst other things: pre-operative assessment and investigation; understanding of normal and abnormal states relating to surgical procedures; identification and treatment of common and important complications; patient management through role-playing, with professional actors playing patients; and carrying out clinical procedures, including taking patient histories, ordering tests, taking blood and putting up intravenous infusions (Wiredu, 2005, p. 114).

6.4   Direct observation and indirect control

The purpose of the PDA-based system, which reports to a centralised database in London, is to ensure that situated learning by each of the medical professionals based around the country can be documented and subjected to assessment by the person responsible. However, this aim is not succeeding. The problems of using the PDA system for this purpose are far beyond usability problems of artificially interjecting the technology into busy hospital work. The aim for strong local control over activities clashes in territorial disputes with the attempts to exercise equally strong remote control from the central learning centre. As a result, the only useful aspects of the PDA have ended up being the individual use of medical dictionaries and the personal information management functionality.

Yin is trapped between two regimes of control competing against each other. The local staff at her hospital where she spends most of her time have ample opportunities to exercise control through direct observation of the work conducted as well as indirect control, for example, through organising the activities they are expected to engage in. The central co-ordinator in London is responsible for monitoring that her learning progresses and seeks to enforce the continual documentation of activities through the uploading of structured forms and reports from her PDA. In this case, local direct observation has the upper hand (Wiredu and Sørensen, 2006). Yin, however, appropriates her use of the PDA from one documenting learning to one serving individual needs for information management, for example, as a diary, providing to-do lists, etc. (Wiredu, 2007).

Yates (1989) analyses how the increase in enterprise size and geographical distribution of organisations from the 1850s to the 1920s represented a significant problem of maintaining operational control. Here formal organisational and technological means of communication and control were established both to co-ordinate distributed activities, to disseminate orders and best practices from the top to each part of the organisation and to systematically collect performance data and pass it up in the organisation. These formal means of exercising control through communication using organisational procedures, plans, schedules, telegrams, forms, etc., replaced the previously informal means of control.

The introduction of flexible and mobile working represents a further source of complexity potentially challenging established regimes of control. Mobile working can increase performance ambiguity if mobilising work reduces the ability to exercise direct behavioural control (Wiredu and Sørensen, 2006). However, indirectly controlling mobile work activities, for example, through monitoring and controlling the outcome, can replace the need for direct control. For work with a high degree of individual discretion, direct observation will be both more difficult and less essential. Here resolving goal incongruence through outcome control is more important. Conversely, for mobile work with very low degrees of discretion, it may be relatively easy to implement forms of mediated direct observation – or surveillance.

6.5   Organising mobility

Whereas the use of mobile services is essential to the cases discussed in this book, there are clearly differences across organisations in terms of generating organisational value from enterprise mobility (Brodt and Verburg, 2007). Revisiting two of the cases can illustrate this: case 3 (mobile foreign exchange trading); and case 2 (London Black Cab work). These two cases illustrate two organising possibilities of enterprise mobility. In the organisation of 24-hour foreign exchange trading, the solution is to mobilise traders in flexible arrangements. The case of London Black Cab work illustrates the opposite aim of organising independent activities.

Organising 24-hour trading

Before settling on mobile trading, the Middle-Eastern bank has over the past three decades implemented three other solutions to the problem of the bank maintaining 24-hour foreign exchange trading (see case 3). Figure 6.1 illustrates the four different solutions in terms of the degree of constant connectivity with the market and on- or off-premises trading. The experiences highlight the importance of contextually balancing both emerging and planned technology performances, remote control and local discretion, and direct observation and indirect control.

24-hour trading floor: in the 1980s the bank extended normal trading hours in the central head-office dealing room in Bahrain to three shifts engaged in 24-hour trading throughout the week. This facilitated constant connectivity with the market, but was also a very expensive solution, as it required many more traders. Furthermore, the solution did not suit the traders particularly well as they were forced to work in shifts.

Home-based trading: in the 1990s the bank extended normal trading hours through home-based trading for a selected group of traders using desktop

Figure 6.1 Four different 24-hour trading solutions adopted by the Middle-Eastern bank in case 3 characterised in terms of their spatial layout (on- or off-premises) and the support for constant connectivity between the trader and the foreign exchange market. Adapted from Sørensen and Al-Taitoon, 2008, p. 923

computers and fixed-line telephone connectivity. This solution was much less expensive for the bank, yet maintained some degree of continuous connection with the market outside of normal trading hours. However, for the traders this solution exerted an excessive burden on their family life by almost permanently chaining them to their homes.

Multi-branch trading rooms: in 1993 the bank engaged a third approach to 24-hour trading by globalising foreign exchange trading across four time zones in dealing rooms located in Bahrain, London, New York and Singapore. This pilot project aimed to solve the problem by geographically distributing trading to follow daytime across the globe. The pilot project was unsuccessful as trading relies extensively on both social networking and the tight co-ordination of activities to engage effectively with the market. Both the management of client relationships and deals turned out to be more complex than originally assumed and were therefore difficult to virtualise. The organisational politics regarding ownership of the client relationship was also a significant barrier to this solution.

Mobile trading: mobile trading was first introduced in the bank in 1997 and is presented and discussed in case 3. In a similar fashion to home-based trading, mobile trading assumes information services innovation as a means of augmenting trader connectivity with the market. The significant difference is the duality of constant market connectivity and geographical mobility.

Organising Black Cab work

The previous example illustrates how organisational choices of loose or tight coupling between traders, the market and the location of trading can render an existing organisation more flexible through enterprise mobility. This is probably a common scenario as traditional organisations seek to engage in arranging work flexibly through re-defining existing barriers (Elaluf-Calderwood, 2008). This section revisits the case of London taxicab drivers (case 2) in order to illustrate the opposite scenario of an organisation that is traditionally comprised of highly independent and mobile individuals who through enterprise mobility can establish a more co-ordinated response to competition.

A licensed London Black Cab driver has traditionally always had extensive independence and is only bound by his or her own decisions concerning working hours and choice of where to position the cab within central London. The introduction of private hire vehicles, or minicabs, to London in 1998 created a competitor to the taxicab. Minicabs are not allowed to pick customers up in the street or wait at taxi ranks but must be called directly by the customer. Minicab drivers do not need to acquire 'the Knowledge', but with a valid driver's licence and a clean bill of health, they can apply for a minicab license and start driving immediately. Also, whereas only a few expensive makes and models of taxicabs are allowed, there are much less strict requirements for minicabs. The much lower barrier to entry has resulted in London taxicab drivers facing a significant competitive challenge. Large minicab companies have emerged, offering more affordable and tightly co-ordinated services to clients and thereby obtaining lucrative long-term corporate contracts.

The uptake of computer-cab systems has been much slower in London than in other countries as taxicab drivers here traditionally have worked on their own supported by 'the Knowledge' (Skok and Baird, 2005). Minicab companies rely critically on dispatch technology, as they are not allowed to approach customers directly in the street. Through enterprise mobility, the minicab companies have mounted significant competitive pressure on the independent London Black Cab drivers, who in turn are increasingly forced to join computer-cab companies in order to respond to this pressure.

The specific technological stipulation of the interdependencies between drivers and the different computer-cab companies is one of the main issues of discussion amongst taxicab drivers – for example, in terms of the relative fairness of job allocation and the requirements on quotas of fares accepted through the system (Elaluf-Calderwood, 2008).

Balancing flexibility and control

The transition from a large group of highly independent taxicab drivers to an organised capability exemplifies the emergence of loose organisation out of independence. Here boundaries are cultivated in terms of the creation of interdependencies and the control of behaviour. Khalid's case (case 3) illustrates the challenge of reducing interdependencies and loosening control. In this case, enterprise mobility is an organisational solution to the problem of balancing flexibility of work and constant connectivity to the market. It is the only one of the four solutions establishing trade-offs acceptable to both the bank and the traders. In case 2 involving Ray, extensive flexibility hinders organised response. Here enterprise mobility provides the opportunity for loosely coupled drivers to organise a response to the market demands whilst retaining a significant element of driver discretion.

6.6   Trust and enterprise mobility

Trust can been defined as 'a psychological state comprising the intention to accept vulnerability based on positive expectations of the intentions or behaviour of another, irrespective of the ability to monitor or control that other party' (Rousseau *et al.*, 1998). Mobile work is both remotely distributed and locally mobile. This implies that the awareness of activities and the co-ordination of interdependencies rely on technology performances. Trust shapes and is shaped by mobile work and mobile technology performances. As argued by Olson and Olson (2000), *distance matters* and working together remotely requires common ground and the pre-disposition for working together across contexts. The reliance on individual discretion implies the requirement of interpersonal trust between the participants and mutual trust between mobile workers and the organisation they work for.

In the context of interpersonal trust at work, a colleague is generally perceived as trustworthy if he or she delivers on work commitments. This assumption of trust, the perception of trustworthiness and the perception of delivery on promises as a result of the unfolding collaboration has been questioned in cases of highly distributed cross-functional work (Zolin *et al.*, 2004). Here trust, perceived trustworthiness and perceived follow-through have been found to be stable over time, indicating that initial impressions last when it is difficult to directly monitor colleagues. However, Dubé and Robey (2009) argue against this proposition and suggest that trustworthiness in remote collaboration is judged by monitoring previous performance and that initial distrust is therefore instrumental in establishing trust. Mobile services play an essential role in monitoring remote performances and, as such, also for interpersonal trust in mobile work. However, the extensive use of technology monitoring mobile performance requires that the person being monitored trusts both colleagues and the organisation.

Mobile services offer possibilities for intimacy, connectivity, memory and pervasiveness. Monitoring mobile work performance can therefore render traces of detailed activities remotely visible for inspection and further processing, facilitating the distributed co-ordination of mutual interdependencies, remote intervention and general sharing of knowledge across mobile workers. Here trust is critical, as this kind of intimate data can be put to a range of uses.

The two extremes of discretion at work can both imply less problematic collection of data than work with some degree of discretion. Simon's experiences as a security guard may indeed confirm the old adage that trust is good but control is better. The RFID-based system represents brute force in completely routinising standard security guard work by directly reporting his position to a central database, which then reports back to Simon where to go next. He can relatively easily assess the trustworthiness of the system, in terms of its direct operations, through his interaction with it. However, he will need to trust the system in terms of what other auxiliary functions it can perform based on the data collected through use.

GPS tracking and mobile email implemented within a group of Danish independently owned long-distance lorries illustrate the case of a high degree of discretion. The GPS system enabled a central shipping agent to observe the exact position and movement of each of the lorries in the fleet (Herskind, 1996). As drivers were all independent contractors, there were no issues of trust – work was managed exclusively by outcome. Lorries were always fully loaded with fresh and salted fish from Norway and Denmark in order to travel to the south of Europe, but the challenge was to respond to available return loads from or within France, Germany, Italy, Spain, etc., to Denmark, Norway or Sweden. Being monitored was likely to bring more work, as the shipping agent's location would help him or her to find return loads. The situation for Ray, the cabbie, and John and Mary, the two police officers, is similar to the Danish lorry drivers in that they stand to directly benefit from allowing their mobile activities to be recorded. Ray will stand a better chance of being offered a nearby job if the computer-cab dispatch office knows where he is and if he is actively looking for a customer. Similarly, the police officers will be able to rapidly receive assistance from colleagues in case it is needed or indeed from an ambulance in case of injury if the control room knows where they are.

However, the British Airways check-in staff at London Heathrow Airport whose strike in 2003 resulted in 80,000 stranded passengers did not trust their management, who claimed that a new system recording location would not lead to micro-management of shifts; nor did they discern any possible benefit from the system (BBC News, 2003). The collection and use of traces of mobile activities represent computational representations of that person, and the readiness to have essential information recorded will depend on both trust and the individual's interests in being represented.

Recording and representing mobile information raises a series of more general social issues of privacy, for example, the potential for RFID technology to generate data on individual consumer behaviour (Albrecht and McIntyre, 2006), the ability of ubiquitous technology to allow consumers to re-balance the surveillance they are subjected to (Mann and Niedzviecki, 2002) and the role of individual information for identity systems (Lyon, 2009; Whitley and Hosein, 2010). Interestingly, this debate is conducted in a context of rapid technological change of location-based social networking gaining widespread popularity.[14]

6.7   Cultivating boundaries

The organisation of mobile working is governed by tensions, paradoxes and contradictory requirements – for example, the requirement of structure in order to obtain flexible teamworking (Dubé and Robey, 2009). This section explores enterprise mobility as the organisation of technology performances enabling both the organisational flow of decisions and the cultivation of control boundaries.

Mobile information supply chains

Mobile working relies on emerging and planned technology performances, which in turn result in flows of messages, decisions, symbols, signals, notifications and other informational objects. Mobile services, amongst others, allow the radical fluidisation of interaction to emerge out of situations. Participants can at the spur of the moment establish connections, request information, publish photos, engage in ongoing conversations, etc. For the participants, variation is an integral aspect of such interaction and flows, rather than simply the connections in networks, can represent the interaction (Mol and Law, 1994). The imagery intended is one of closely integrating and orchestrating heterogeneous actors in a smooth flow of information, communication and decisions with blurring boundaries across participating organisational units and individuals.

As with the desire for individuals to be able to engage in fluid interactivity, this can also be understood at the aggregate level of teams, supply chains, crowds, organisations and networks of partners aiming for the interaction to be experienced as a smooth flow of interaction and decisions. Terms such as information supply chains (March *et al.*, 2008), knowledge supply chains (Choi *et al.*, 2004; Cha *et al.*, 2008) and innovation supply chains (Desbarats, 1999) have been applied to explain flexible collaboration in information work. A *mobile information supply chain* is a concept that borrows from the characterisation of manufacturing processes as supply chains. This concept embodies assumptions of both flexibility and structure. It represents structure without rigidity and fluidity without purposeless chaos.

The mobile information supply chain offers the attractive notion that the challenges of managing mobile work are similar to those of managing manufacturing supply chains. However, whereas the technology supporting the management of manufacturing supply chains can freely record essential information about the industrial processes engaging physical components, the mobile information supply chain is not made up of objects and industrial processes but of human information workers and information management processes. Unlike their industrial components, these mobile information workers will be keen to understand what is being written about them in representations of the information supply chain and they will also have strong opinions of how this information is applied.

The British Airways check-in personnel showed this clearly when they refused to work with a new system that recorded their behaviour (BBC News, 2003).

However, the emphasis on decision flows in mobile information supply chains does emphasise the need for the explicit management and planning of activities – planned organisational interventions through planned technological performances. Mobile information supply chains consist of combinations of restricted and unrestricted information flows, and the precise arrangement of the restrictions defines planned boundaries for communication and information flows.

Balancing fluidity and boundaries

Carefully cultivated mobile information supply chains can achieve the desired balance between restricted and unrestricted flows and function as designed boundaries within or between organisational units. The main challenge is to implement decision flows that provide both guidance and structure while remaining flexible and malleable (Schmidt and Simone, 1996). An unrestricted flow of information across boundaries can support emerging connectivity shaped by organisational contingencies calling for emerging decisions. This in turn can lead to innovative decisions re-combining existing organisational resources in novel and unexpected ways (Ciborra, 2002, Chapter 3). However, such an unrestricted flow of information comes at a price in terms of the individual need to manage the flow on a day-to-day basis. This can result in extensive unplanned meetings, constant interruptions and working days filled with micro-co-ordination. In this sense, the organisation of mobile information supply chains represents the formalisation of the co-ordination and control of mutual interdependencies, and this formalisation reduces the complexity of negotiating these interdependencies as it has been organised through the supply chain. The activities of cultivating the balance between emerging and planned technology performances relate to the everyday work of standardisation within organisations (Hughes *et al.*, 2002).

Symmetry and asymmetry in interaction flows

The contextual balancing of mobile information supply chains signifies the appropriation of technology symmetry and asymmetry. The organisation of flows requires some degree of prioritisation either in the form of technology symmetry socially appropriated with prioritisation or directly through technology asymmetry. Planned technology performances stipulating interaction asymmetry through priorities serve as technologically embedded rules governing acceptable interaction patterns. Such organisation of interaction asymmetry will be supplemented by traditional organisational measures, such as collective practices, informal rules and standard operating procedures. The ensemble of socio-technical measures can both reduce the complexity and uncertainty of interaction and strengthen the boundaries between parts of the mobile information supply chain.

Considering field-force management as an example, so far we have considered the cases of a security guard (case 5), an industrial waste lorry driver (case 6) and a food delivery driver (case 8). In case 8, the organisation aimed at designing a flow of information where both delivery drivers and customers call the contact centre directly in case of problems, emerging queries, etc. However, such prioritisation was only implemented through stipulated practices and, as the delivery drivers return to the same group of customers on a regular basis, a counter-flow emerges when drivers provide the customers with their mobile phone numbers. In the case of Tesco.com home-delivery drivers, customers are required to provide a telephone number at which they can be contacted by the delivery driver. However, the customer cannot directly contact the delivery driver to inquire, for example, about the exact time of delivery, as each weekly delivery is made by different drivers. Even in the cases where the driver may call the customer, the caller ID is blocked. As in case 8, the aim is to ensure that the customer calls the central customer contact centre, where an agent will then call the specific store from which the delivery is being sent. Here the store's delivery manager will call the delivery driver's mobile phone and, if all goes well, the delivery driver will call the customer. The asymmetry is simple to see, with an asymmetric information flow providing a one-step connection between the driver and the customer in one direction and in the reverse direction a three-step connection between the customer and the driver. This mobile information supply chain is characterised by interaction asymmetry jointly defined by organisational practices but also stipulated through mobile co-ordination mechanisms.

Influence, control and asymmetry

Different participants will naturally exercise different amounts of influence and control in the ongoing cultivation of the mobile information flows. By deploying systems, the organisation can seek to enforce specific control boundaries, as the Tesco.com example illustrates. An individual with significant organisational influence, such as Hiro the CEO, has extensive powers to determine the mobile information supply chain, organising every request into an email. In the hands of the 'interactional upper class', mobile technology can become a rich instrument for remote shaping. For others, it can conversely signify a reduced ability to orchestrate even the immediate work process, as these can be subjected to remote scrutiny. Jason, the delivery driver in case 8, is constantly engaged in the ongoing negotiation of organisational boundaries when what to him seems practical in the situation is to the organisation a violation of the principles for interaction.

The organisational boundaries cultivated through mobile information flows will critically depend on the specific context in which they are enacted and enforced. In some organisational contexts there will be significant emphasis on the extensive sharing of information, with participants exercising considerable discretion regarding how this information is used, while in other contexts information flows will be carefully designed and cultivated to best match specific business processes. However, the specific organisational politics of information sharing can of course greatly influence the feasibility of unbounded interaction. The example of a product development group in a European country using a collaborative system to document their design process offers a good insight into this (Ciborra and Patriotta, 1996). The developers were happily going about their work until one day the CEO of the corporation, based in London, posted a message congratulating their efforts and applauding their progress. As none of the participants had considered the possibility that they were being observed from above, they were quite nervous about the prospects of having all their detailed internal discussions observed externally through the groupware system. Consequently they immediately stopped using the system and could only be persuaded to do so when the organisation ensured that access to the system replicated the organisational arrangements where only people who were an integral and trusted part of the information flows had access to view the detailed data in the system.

6.8   Summary

This chapter has explored the role of mobile technology performances in exercising and being subjected to control. The two cases of a food delivery driver and a healthcare professional further add to the discussion of enterprise mobility, and these two cases raise particular issues concerning control. The first case exemplifies the tension between remote ability to control and the degree of retained local discretion. The second case demonstrates the possibility of conflict between direct localised observation and attempts to engage in indirect control through planned technology performances. Enterprise mobility can both render tightly organised work flexible and, conversely, support the tighter organisation of loosely connected activities. The case of taxicab work illustrates the latter. Here work that is traditionally highly individual and independent of the activities of others can become organised through mobile interconnectivity. Interpersonal trust and trustworthiness as well as mutual trust between the organisation and the mobile worker play a key role in ensuring a high degree of sophistication in the use of mobile services. This chapter has synthesised the discussion of control into the concept of mobile information supply chains, highlighting the contextual balancing of stipulated performances of prioritised interaction with support for flexible adaptation of the information flows.

7
Portfolios – Amplified Mobility

This chapter highlights the diversity of mobile services explored from the nine cases of enterprise mobility in order to discuss mobile service portfolios and the materiality of mobile services. It emphasises the importance of understanding the diversity of mobile services in service portfolios. Such diversity provides the platform for innovative technology performances in mobile work.

7.1   Ecologies, infrastructures and portfolios

The technological development from isolated, almost single-purpose main-frames to a diversity of capabilities and services places great emphasis on the understanding of this diversity of services (Mathiassen and Sørensen, 2008). A variety of portfolio configurations may be needed for different mobile work roles or over time for the same role (Burton-Jones and Gallivan, 2007). Furthermore, technological development provides constant innovation, expanding diversity, and research must understand the significant changes in such development. The challenge is to characterise general categories that will be stable over a period of technological innovation and not merely to categorise the current state of the art. Leifer's (1988) categorisation of the diversity of technological provision is now no longer a good representation of the diversity of technology as it is based on the mainframe versus personal computer technology of the 1980s.

Adomavicius *et al.* (2007 and 2008) suggest understanding rapid techno-logical innovation by mapping the information technology landscape using a ecosystem concept emphasising the dynamics, i.e.: 'A system of interrelated technologies that influence each other's evolution and development. A specific technology ecosystem view is defined around a focal technology in a given context' (Adomavicius *et al.*, 2007, p. 201). This perspective is mainly concerned with innovation and the interrelation of components and systems. The concept of modularity can support the exploration of product and service architectures (Baldwin and Clark, 2000) – for example, characterising the modularity of services in three layers of the supply chain, bundles and components (Voss and Hsuan, 2009). Yoo *et al.* (2010) argue that the convergence of physical products and software implies a need to consider hybrid architectures characterised by both modularity and layers.

The gradual shift during the past half-decade from one unconnected main-frame to rich ecosystems of easily re-combinable technologies and services has not significantly informed the conceptualisation of information services at work. However, notable exceptions are the discussions of infrastructures for action and experience (Star and Ruhleder, 1996; Dourish and Bell, 2007), infor-mation ecologies (Nardi and O'Day, 1999) and portfolios of services (Broadbent *et al.*, 1999; Peppard, 2003; Mathiassen and Sørensen, 2008).

Susskind (2008), for example, characterises the diversity of information services within the legal sector by distinguishing between client- and internal-facing serv-ices, and between operational and knowledge processes. This yields the following four types: back-office technology; client relationship systems; internal knowledge systems; and on-line legal services. Mathiassen and Sørensen (2008) offer a discus-sion emphasising the diversity of general organisational information services sup-porting decision-making under varying degrees of equivocality and uncertainty in terms of: computational services standardising algorithmic encounters; network-ing services standardising connections; adaptive services standardising informa-tion; and collaborative services standardising the material for collaboration.

7.2   Mobile services diversity

Within the discussion of enterprise mobility portfolios, there are pragmatic sug-gestions for the assembly of a suitable enterprise mobility service portfolio (see, for example, Hayes and Kuchinskas, 2003; Lattanzi *et al.*, 2006), as well as a vari-ety of academic and practitioner reflections on the enterprise mobility ecosystem (Basole, 2009). Jarvenpaa and Lang's (2005) suggestion for system design options in order to balance paradoxes is an example of portfolio diversity. They suggest a range of design features supporting the users: presence management, collabo-ration support, context-awareness, location awareness and role management. These are all aspects related directly to services beyond simple connectivity and particularly emphasise ongoing relationships as opposed to encounters. Presence, location and role management relate to the intimacy of services by associating assumptions about the user with services. Context-awareness relates to the perva-siveness of services. An information services portfolio supporting mobile technol-ogy performances is in this book defined as a modular collection of enterprise mobility services as outlined in Chapter 2 supporting (see Figure 7.1):

- *Intimacy* – either distant and anonymous services or support for bodily close-ness of user-service interaction with the possibility of an identifiable user.

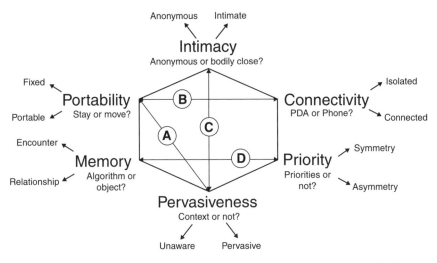

Figure 7.1 Mobile services diversity containing six categories of services and with four pairs of services highlighted

- *Connectivity* – isolated service or the support for easy remote connectivity with other users and services.
- *Pervasiveness* – with services embedding contextual information through awareness of the environment.
- *Memory* – through the support for transactions without memory or ongoing relationships with memory.
- *Priority* – unprioritised interaction symmetry or prioritisation of interaction through embedded asymmetry.
- *Portability* – the ability for the user to carry with him or her devices offering instant access to services.

Service diversity in action

These six categories of services each represent a cluster of services to be defined first by developers, then instantiated and personalised by a mobile worker, in a similar fashion to the definition of service categories in Mathiassen and Sørensen (2008). Enterprise mobility services are here considered to be configural, i.e., heterogeneous use emerging as a distinct pattern, rather than shared, where individuals, groups and organisations adopt homogeneous patterns of use (Burton-Jones and Gallivan, 2007, p. 666). Across the nine cases discussed throughout the book, the six categories of services are used in diverse manners and to varying extents. Figure 7.2 illustrates the notation used to characterise this diversity through the simple example of standard mobile connectivity.

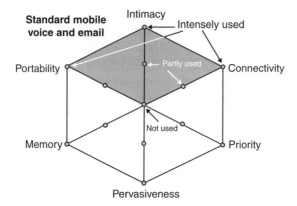

Figure 7.2 Example explaining the principle of mapping technology performances across the six categories of services

Figure 7.3 shows the diversity of services across the cases. While this is a simple graphical illustration of complex technology performances, it demonstrates a diversity of forms of reliance on mobile services beyond the most commonly discussed types of portable connectivity.

The diverse reliance on mobile services in Figure 7.3 shows light reliance on mobile interaction services beyond direct connectivity in the cases of the food delivery driver (Jason – case 8), the cab driver (Ray – case 2) and the town planner (Jun – case 7). In addition to connectivity, Ray also relies on priority and pervasiveness through the dedicated computer-cab system. Hiro, the CEO, and Khalid, the trader, both rely heavily on memory and priority services in order to support their cultivation of interaction. In the case of Jin, the aim of relying on an ongoing intimate mobile collection of experiences failed, and instead practices rely on intimate portability with access to personal information management and medical reference databases. John and Mary, the police officers, are partly vehicle-bound. This implies less continual reliance on portability, even if they use their shoulder-mounted radio and their mobile phones. Continuous streams of updates from colleagues and the control room offer a degree of pervasiveness providing essential awareness. The two RFID cases of Simon and Winters illustrate a broad reliance on all service aspects in close coupling between mobile services offering: memory of an ongoing work process; pervasiveness through directed interaction depending on data recordings; and prioritised interaction between the mobile worker and a central system.

7.3   Unpacking mobile service diversity

The six dimensions of enterprise mobility services can, for example, be combined into 15 unique pairs of services, but considering that each dimension is

145

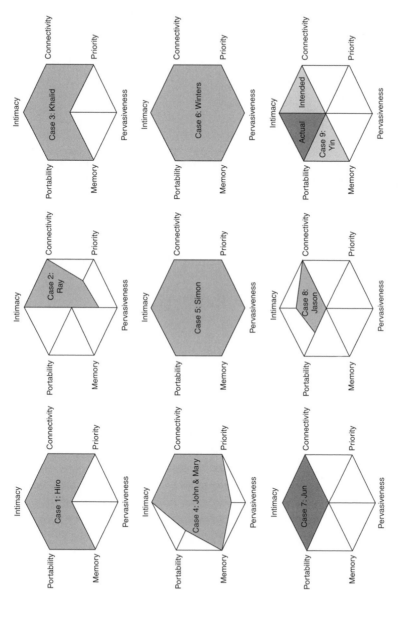

Figure 7.3 Portfolio diversity of services across the nine cases discussed throughout this book

defined in terms of a pair of opposites, there are 60 different combinations of service elements.[15] In the context of this analysis, four pairs of service dimensions will be explored as an illustration of the diversity mobile services – the four highlighted connections in Figure 7.1 labelled A, B, C and D. These four combinations of the 15 service pairs are illustrated in Figure 7.4: A) portable pervasiveness; B) connected portability; C) user-environment symbiosis; and D) relationship priority. The following text discusses these four service pairs in order to unpack the design choices associated with the diversity. Each pair of service categories discussed is characterised by four archetypical service element choices, yielding a total of 16 service elements in this selection. The names of these are pragmatically chosen without further theoretical considerations for any possible overlap in categories. The formalisation here assumes that a *node*, which is portable and not wirelessly connected, is distinct from a *device*, which is also stationary and in addition is unaware of its context. For example, it has been assumed that a portable and unconnected service is distinct from a *utility* or *intimant*, which is unaware of its environment, and may or may not be portable (see Figure 7.4). As a result of this assumption, creative naming has been applied in order to span the 16 combinations discussed. In these service combinations, each type is merely a building block for comprehensive services.

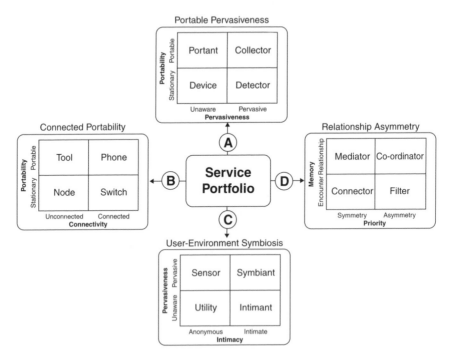

Figure 7.4   Diversity of ubiquity, mobility, intimacy and priority services

They are similar in type to the four basic organisational information services presented in Mathiassen and Sørensen (2008).

Aspect A = portable pervasiveness

Combining choices between portable and stationary mobile services with those of pervasive services as opposed to those unaware of the immediate environment results in a service diversity spanning; the *device*, stationary and unaware; the *portant* – portable, yet contextually unaware; the *detector*, reflecting stationary awareness of environmental aspects; and the *collector*, combining portable and pervasive services. Adding the third service category of connectivity yields Lyytinen and Yoo's (2002a) definition of ubiquity as mobile (connected portability) services, which are pervasive. The RFID-enabled mobile phone used in the two RFID cases is in terms of this categorisation an example of a connected collector.

Aspect B = connected portability

The combination of the service categories portability and connectivity signifies the diversity of options in establishing portable connectivity spanning: the stationary-unconnected *node*; the *switch* providing stationary connectivity; the portable but unconnected *tool*; and the *phone*, which here is the shorthand for portable connectivity and is by far the most common type of service. Indeed, in all the cases, some form of portable connectivity is provided through mobile phones. However, for the vehicle-based work in cases 2, 4, 5, 6 and 8, portability is not always necessary or even desirable, yet vehicle-based services can still be an integrated part of mobile working. The PDA in case 9 was originally intended to serve as a means of establishing portable connections but, through circumstances and Yin's decisions, turned into a portable but unconnected *tool*. In the case of Simon (case 5), the torch previously used to collect data for weekly uploads can be characterised as a *collector tool*, which in turn was replaced by the RFID reader-enabled mobile phone, i.e., the *collector phone*.

Aspect C = user-environment symbiosis

Combining services for user-intimacy and pervasiveness explores the diversity of services forging user-environment symbiosis. User intimacy here implies the extent to which services are either anonymous or relate directly to the individual user, for example, in terms of the closeness of the coupling between the service and the user's identity. Similarly, pervasiveness denotes the extent to which the service is either unaware of its environment or relies significantly on specific interactions with environmental awareness. The diversity here spans: the *utility*, with no direct relations to either the specific user or the immediate environment; the anonymous yet pervasive *sensor*; the personalised yet environmentally unaware intimant; and the *symbiant*, which synthesises user intimacy and environment awareness.

The case studies provide several examples of service intimacy through services associated with individuals, for example, Hiro and Khalid's personalisation of their phone and SmartWatch – *intimant phones*. Indeed, all cases demonstrated some extent of intimacy, which is not abnormal, as service intimacy is a core aspect of enterprise mobility. There is, however, less comprehensive representation of pervasive services in the portfolios. The RFID portfolio in cases 5 and 6 offered the best example of extensive application of a pervasive service, and the combination of a highly personal device swiping embedded RFID tags exemplifies a *symbiant collector*.

Aspect D = relationship asymmetry

The type of mediation and the extent to which mediation employs embedded priority of the interaction form a core aspect of the support for mobile workers. Supporting the diversity of mediation in the form of encounters and relationships, as well as the diversity of technology-supported priorities, for example, symmetry and asymmetry, ensures the ability to contextually balance between flexible connectivity and ongoing mediated discussions. The *connector* offers the flexible possibility of encounter symmetry, the *filter* prioritises encounters, the *mediator* supports unprioritised ongoing relationships and the *co-ordinator* provides opportunities for prioritised relationships.

These four types represent a diversity of affordances, which will be reflected in a diversity of technology performances depending on user preferences, needs and the particular situation. The four types can each influence the extent to which the individual can cultivate interaction asymmetry to best suit their perceived needs. Figure 7.5 illustrates the four categories of affordances through general examples, while Figure 7.6 uses examples from the case studies, with the number indicating the case number.

The successful cultivation of priority relies critically on the specific context. Simon, the security guard, requires the explicit mediation of ongoing relationships through mediator or co-ordinator mechanisms ensuring sufficient technological support. However, Ray, the Black Cab driver, mainly requires a relatively simple connector mechanism to ensure the appropriate levels of individual discretion for the user. This is supplemented with the computer-cab system functioning as a co-ordinator.

Table 7.1 illustrates at a more detailed level than Figure 7.3 the diversity of services through examples from the cases using the classification of diversity suggested in Figure 7.4.

7.4   Affordances, mechanisms and materiality

The aim of this book is to explore mobile service diversity. The provision of services is here used as shorthand for underlying assemblages of affordances,

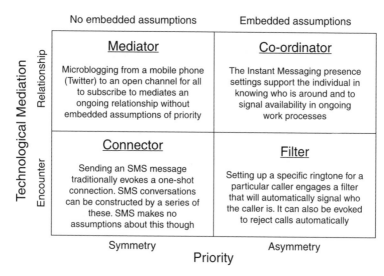

Figure 7.5 Diversity of priority services exemplified

constraints and mechanisms. Affordances are conceptualised as the technological provision of resources for reflexive action by the user but are also associated with user and context assumptions about usefulness (Gaver, 1991). Affordances contain in their essence a socio-material relationship. The six categories of

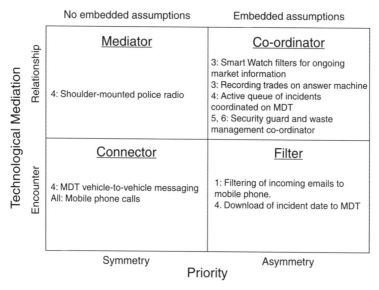

Figure 7.6 Examples of priority services from the case studies. The item numbers indicate the case study number

Table 7.1 Characterisation of service diversity with examples drawn from the general use of mobile phones and from the case studies

Case	Name	Job	Service	Portfolio categorisation			
				A	B	C	D
–	–	–	Mobile call and SMS	Portant	Phone	Intimant	Connector
–	–	–	iPhone Google Maps	Collector	Phone	Symbiant	Co-ordinator
1	Hiro	CEO	Mobile phone	Portant	Phone	Intimant	Co-ordinator
2	Ray	Cabbie	Computer-cab system	Detector	Switch	Symbiant	Co-ordinator
2	Ray	Cabbie	Mobile auto-match	Device	Phone	Intimant	Connector
3	Khalid	Trader	SmartWatch	Portant	Phone	Intimant	Co-ordinator
3	Khalid	Trader	Answer machine	Device	Switch	Utility	Co-ordinator
4	John and Mary	Police	Mobile Data Terminal	Device	Switch	Utility	Co-ordinator
4	John and Mary	Police	Personal radio	Portant	Phone	Intimant	Mediator
6	Simon	Security	Torch	Collector	Tool	Symbiant	Co-ordinator
6	Simon	Security	RFID phone	Collector	Phone	Symbiant	Co-ordinator
9	Yin	PSP	PDA (intended)	Portant	Phone	Intimant	Co-ordinator
9	Yin	PSP	PDA (as used)	Portant	Tool	Intimant	Co-ordinator

services outlined (see Figure 7.1) denote categories of services, which embed specific assumptions about the user-context-service relationship. As interactive objects, an extensive part of these assumptions will be cultivated through use. Constraints relate directly to affordances – for example, in the lack of connectivity – or other aspects constraining the access to a particular affordance. Mechanisms are here conceptualised as formal constructs either acting as cognitive maps informing the user of possible actions or as scripts stipulating specific courses of action (Schmidt, 1999).

Co-ordination mechanisms and the granularity of services

Given the primary focus here on interaction, the use of mechanisms as formal constructs for the co-ordination of distributed mobile activities is particularly interesting. Such co-ordination mechanisms can, for example, be passive and offer conceptual structures (Carstensen and Sørensen, 1996) or, as interactive computational mechanisms, offer more direct assistence in the support for distributed co-ordination of activities (Schmidt and Simone, 1996). The discussion of co-ordination mechanisms can be conducted at the three different analytical levels of organisations, groups and individuals.

At the organisational level, before the 1980s the co-ordination mechanism was seen as situated at the aggregate organisational level and was traditionally implicitly assumed to be technological support with little or no discussion of materiality or technology (see, for example, Thompson, 1967; Van de Ven *et al.*, 1976; Martinez and Jarillo, 1989). The granularity of co-ordination mechanisms is here one of the organisational unit enabling a discussion of interdependencies between such units and how the firm as a whole represents a co-ordination mechanism. This view of the co-ordination mechanism coincided with mainframe technology, providing one point of access for the computerised support of co-ordination.

The discussion of co-ordination mechanisms as formal constructs supporting horizontal co-ordination of mutual interdependencies between peer groups emerged as part of the groupware and CSCW discourse in the mid-1980s (see, for example, Malone *et al.*, 1987; Schmidt and Bannon, 1992; Malone and Crowston, 2001). Here the technological provision of networked personal computers along with increasingly distributed collaboration implied a finer granularity of co-ordination (Grandori, 1997) in performance teams (Goffman, 1959, Chapter 2) and an increased reliance on information technology in the co-ordination of these distributed activities.

With the further fragmentation and distribution of work activities along with the widespread diffusion of mobile information technology supported by global infrastructures, the co-ordination mechanism has increasingly become individualised. The individualisation and intensification combined with readily available mobile interaction support both myriads of emergent connections as well as the placement of organisational and team co-ordination mechanisms in the hands of individual mobile workers.

Computational co-ordination mechanisms

Primarily distinguishing between synchronous and asynchronous interaction modalities, Van Fenema and Kotlarsky (2008) characterise information technology supporting personal, impersonal and automated co-ordination. Within the context of the assumptions in this book, it is not possible to *a priori* determine to what extent a co-ordination mechanism provides an impersonal means of co-ordination if it automates the co-ordination. This relates directly to the discussion of the extent to which material artifacts merely provide resources for decisions or indeed stipulate action (Schmidt, 1999).

The materiality of mechanisms supporting or stipulating the co-ordination of activities can be characterised as a protocol embedded within an artifact (Schmidt and Simone, 1996). It can formalise aspects of both the common field of work, such as the classification of objects, processes and other resources, and the collaborative work environment, such as actors, tasks and activities (Schmidt and Simone, 1996, p. 190). One of the key requirements is for a co-ordination mechanism to be malleable (Schmidt and Simone, 1996, p. 184) in order to be applicable in slightly different and emergent situations than the particular one that was valid when it was instantiated. This requirement of malleability can be interpreted as the necessary resolution of the duality of organisational life between planned interventions and emergent action.

Materiality

Discussions of affordances and performances, and the role of mechanisms as either maps or scripts, relate to the discussion of materiality in organisations – the co-construction of material and organisational practices (Orlikowski, 2007; Orlikowski and Scott, 2008). The discussion of mobile services and technology performances is an example of mutual shaping of the user and the technology through ongoing practices (Orlikowski and Scott, 2008, p. 438) – Research Stream II, where Research Stream I assumes that the technical and the social are separate discrete entities with inherent characteristics. Although the separation into affordances and performances seemingly treats the elements separately, the contextual definition of affordances as the foundational assumption directly co-defines the social and the technical as latent opportunities. Affordances as opportunities for action are not defined in Gibson's (1977 and 1979) sense as inherent properties of the environment, but as relational in terms of the artifact, actor and context (Gaver, 1991). The distinction between affordances and performances emphasises the distinction between a space of contextual opportunities and the actualisation of these opportunities. It enables a discussion of the diversity of opportunities rather than exclusively of the diversity of practices.

Leonardi and Barley (2010) characterise the social construction of organisational technology in terms of perception, interpretation, appropriation, enactment and alignment. The perspective assumed in this book is concerned with the enactment and alignment phases where mobile services are brought into use and the organisation adapts to the technology (Leonardi and Barley, 2010, p. 7). The discussion of the social construction of technology in the context of enactment and alignment relates to the discussions of materiality as maps for situated action or formally scripted behaviour (Suchman, 1987 and 1994; Schmidt, 1999).

This analysis of mobile services adds to the discussion of materiality in organisations in accounts of the understanding of the diversity of digital interactive work objects subjected to individual appropriation and use, the role of interactivity in the cultivation of digital work objects and the materiality of mobile services.

Digital interactive work objects

A *digital interactive object* is here defined as a specific portfolio of services instantiated by an individual and subjected to appropriation and use.[16] Digital interactive objects denote assumptions of the underlying computational capabilities as interactive objects rather than algorithmic codifications, and emphasise the importance of the shaping of these objects through populating with data and reconfiguring and reshaping in other ways (Wegner, 1997; Goldin *et al.*, 2006). The assumption here is that interactive digital interactive objects play an increasingly important role in organisational decision-making in general and in mobile information work in particular. Through everyday interactions, each individually used mobile phone is appropriated into a unique element in a symbiotic relationship.

Interactive cultivation of digital objects

In terms of *interactivity*, this is concerned both with the processes of cultivating the data within digital interactive objects and with the shifting nature of the distinction between algorithms, data and meta-data. The algorithmic transaction paradigm for computing assumes a relatively stable boundary between what is considered the program (as envisioned by the designer and programmer) and what is considered data to be provided in the context of use. However, processes of abstraction and the flexible design of information services can produce digital interactive objects with greater flexibility. Characteristic for many contemporary computer applications is just such flexibility in the distinction between data and programme/algorithm, where they were initially fixed. For example, traditional mainframe back-office applications provided fixed structural and algorithmic templates for storing data in structured, pre-defined databases and a series of fixed, pre-determined queries to be evoked, each resulting in a report to be printed. The advent of spreadsheets, database systems and loosely structured data-storage applications on the personal computer provides the ability for a much more flexible appropriation by the user. The extensive parametrisation of aspects of applications, such as tailorable interfaces, also provides plasticity in the cultivation of the digital interactive objects. Contemporary smartphones with associated application stores provide a semi-blank canvas with basic properties pre-loaded awaiting an ongoing (lifelong?) process of individualisation and appropriation.

Digital object convergence

Furthermore, the digitisation of interactive objects and digital convergence imply that output from any object can in principle become input to any other process (Tilson *et al.*, 2010). This implies a wealth of opportunities for the novel association, re-combination and aggregation of objects during enactment and alignment. One of the radical changes in the use of personal computer applications was the introduction of the clipboard, which allowed the simple convergence of data between applications so that the spreadsheet numbers could easily be copied and pasted into the word processing application (Laing, 2004). In terms of contemporary digital interactive objects for mobile information work, the potentials are far from realised. The smartphone, with its individual applications, carries the potential to form parts of complex strings of digital interactive objects. The understanding of the provision of the service portfolios for such strings or webs of digital interactive objects requires close attention to be paid to the practices of use and not only to the theoretical opportunities. Studies of enterprise mobility practices such as those presented and discussed in this book can inform such design considerations.

Services and products

So far, the discussion has used the term 'service' as shorthand for the affordances and mechanisms subjected to technology performances by the user, in the same manner as it is used to characterise organisational information services in Mathiassen and Sørensen (2008). This technically contravenes the core definition of a service, i.e., the retention of service ownership by the service provider (Grönroos, 2000; Mathiassen and Sørensen, 2008). By applying this distinction between products and services, the issue of interactivity becomes even more pertinent as it raises essential issues of control. A product can be subjected to everyday appropriation by the user without the knowledge and attention of the provider of the product. When purchasing a new pair of shoes, the owner will spend time wearing them in order to ensure they fit and the company who produced the shoes will not be engaged. A mobile phone user can continuously update the names and phone numbers stored on his or her SIM card without involving the mobile service provider or indeed the mobile phone handset manufacturer. Similarly, most personal computer owners will install a number of applications without the direct involvement of the equipment, operating system

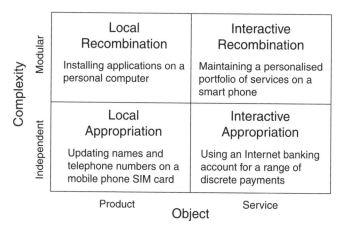

Figure 7.7 The characterisation of the appropriation of independent and modular products and services

or application manufacturer. However, the increasing design of services and the association of services with products necessitate an ongoing services relationship as the services are provided and controlled by the service provider. Considering the distinction between isolated and modular products and services reveals four analytical categories illustrated in Figure 7.7. The local appropriation of names and addresses stored in a mobile phone SIM card differs from the service relationship of changing the name and address in an address book stored on a cloud service such as Google Calendars or .Me, in that the user controls the former and the service provider controls the latter. The distinction between independent and modular products and services signifies the additional problems associated with interdependency between various elements.

Product-service hybrids

Digital convergence produces hybrid forms where it can be difficult to distinguish between products and services, such as in the case of the Amazon Kindle reader, where owners pay for and download e-books. The digital object of the individual e-book may display properties of a product very similar to the standard printed book subjected to individual adaptation through bookmarks and annotations. However, in the case of Orwell's book *1984*, it turned out that Amazon did not have the appropriate permission to sell the book electronically and as a consequence applied the paragraph in the user agreement allowing for books to be revoked. As the Kindle is permanently connected, owners would suddenly, and indeed ironically given the content of the book, see their purchased copy of *1984* disappear as it was remotely deleted from the devices by Amazon (Pogue, 2009). Similarly, most modern mobile phones can be remotely deactivated when stolen or lost.

Service relationships and control

The design of service provision is therefore not merely a matter of providing capabilities exclusively for the individual to appropriate. It denotes an ongoing services relationship where the provider will be much more deeply engaged in the everyday appropriation than with pure products. This implies a much more precarious relationship. The transformation of the traditional provider-user relationship for personal computers demonstrates this. In order to provide improved service (and to safeguard against software piracy), it is often the case that application and operating systems suppliers support an ongoing interactive process of semi-automatic updates of the software and operating system. This can, for example, support the battle against computer viruses and generally ensure that everything is up to date. However, it also creates the ongoing need for the user to engage in interaction with previously non-existing services asking if the user wants the printer drivers and anti-virus software to be updated. The ability to work on an isolated modular system is rapidly disappearing and interactivity is increasingly required as services replace products.

The design of control points in service delivery platforms marks a considerable challenge as it impinges on issues spanning from individual microactivities to issues of business models for revenue generation, infrastructure standards and control (Gawer, 2009; Eaton *et al.*, 2010; Herzhoff *et al.*, 2010). The design considerations will rely on discussions of appropriate metaphors supporting the understanding of digital interactive objects. Are these tools for individual expertise cultivation (Ehn and Kyng, 1985), media supporting discourses (Andersen *et al.*, 1993), language-action systems (Winograd and Flores, 1986), intelligent agents automating activities (Maes, 1991) or infrastructures providing converged capabilities (Ciborra and Associates, 2000)? In the case of products, the distinction between these categories can to a large extent be determined in isolation by the individual user or jointly between a group of collaborators, and can inform the discretionary appropriation of the digital interactive object. However, for services, the specific design of control points by the service provider can inform, steer, enforce and hinder specific interpretations. In terms of informing the further debate of materiality in organisations in general and the provision of enterprise mobility, it can be argued that the study of the diversity, interactivity, convergence and control of digital objects at work can provide further insights into the complex and ever-changing sociomateriality.

7.5 Summary

Much research has studied the social appropriation of mobile connectivity assuming a relatively fixed set of services enabling unprioritised and contextually unaware connections between individuals. Such discussions greatly misrepresent the dynamic character of mobile service innovation and the complexity residing in enterprise mobility portfolios. Whilst portability and connectivity are essential aspects of enterprise mobility, they are not the only ones, and the dual purposes of this book are to both explore the diversity of mobile services at work and to characterise these more generally. This chapter has explored the diversity of mobile services portfolios in terms of the six service categories: intimacy, connectivity, pervasiveness, memory, priority and portability. Four exemplar combinations of two service categories out of the 15 possible have been discussed and related to the case study examples of service portfolio diversity: portable pervasiveness, connected portability, intimate pervasiveness and relationship priority. Such exploration of the diversity of enterprise mobility services can inform the discussion of both designer choices and technology performance choices in the everyday unfolding of mobile work. The discussion of information services diversity is also related to the emerging discussion of materiality and is here informed by the distinction between products and services.

8
Challenges – Managing Mobile Performances

This chapter reviews and discusses the mobile practices explored in Chapters 4, 5 and 6, and discusses more generally the challenges of managing enterprise mobility and the opportunities for both incremental and radical organisational innovation through novel resolution of the fluidity/barrier paradox. Furthermore, it concludes by outlining the research challenges of the broad and sprawling studies of mobilities.

8.1 Emerging and planned performances

The nine cases studies presented and discussed in Chapters 4, 5 and 6 provide a rich variety of enterprise mobility experiences. Table 8.1 characterises the cases using categories from Felstead *et al.* (2005) and Lamond *et al.* (2003) in terms of the type and variety of mobile working arrangements, knowledge intensity, and the extent of internal and external organisational contact. In addition, categories characterise the degree of direct electronic surveillance, discretional decision-making and the requirement of contextual ambidexterity of the mobile worker. All nine cases demonstrate some element of nomadic (Lamond *et al.*, 2003) or mobile working, and in five of the cases multiple locations or clusters. There is a mixture of single and multiple workscapes and workstations, and a variety across all other aspects, except mobile information technology use, which is either high or medium. The two cases (5 and 6) relying on RFID technology demonstrate, in case 5, the shift from a high to a low degree of human intra-organisational contact, being replaced with an increased automated intra- and extra-organisational interaction, and, in case 6, the transformation from a low to a high degree of intra-organisational contact.

As argued in Chapter 1, by far most of the socio-technical research into mobile computing is concerned with the general social appropriation of mobile communication (portable connectivity) and much less attention has been devoted to the use of diverse portfolios of mobile services at work.

Table 8.1 Characterisation of the nine cases using aspects based on categories from Felstead et al. (2005) and Lamond et al. (2003), with the addition of the category of direct electronic surveillance and individual discretion

	1) Hiro CEO	2) Ray Cabbie	3) Khalid Trader	4) John and Mary Police officers	5) Simon Security	6) Winters Lorry driver	7) Jun Planner	8) Jason Lorry driver	9) Yin PSP
Location/cluster 1	Nomadic	Nomadic	Office	Nomadic	Nomadic	Nomadic	Nomadic	Nomadic	Nomadic
Location/cluster 2	Office	–	Nomadic	Office	–	–	Office	–	Remote
Location/cluster 3	Home	–	Home	–	–	–	Home	–	Home
Workscapes	Multiple	Single	Multiple	Multiple	Single	Multiple	Multiple	Single	Multiple
Workstations	Multiple	Single	Multiple	Multiple	Multiple	Single	Multiple	Multiple	Multiple
Multi-functional stations	High	Low	High	Low	Low	Low	Low	Low	High
Discretionary decisions	High	High	High	High	Low	Low	High	Medium	Medium
Work/life boundary	Low	Low	Low	High	High	High	Low	High	High
Personalise work times	High	High	High	Medium	Low	Low	High	Low	Low
Knowledge intensity	High	High	High	Medium	Low	Low	High	Low	High
Ambidexterity	High	High	High	Medium	Low	Low	High	Low	Medium
Intra-org. contact	High	Low	High	High	High=>Low	Low=>High	High	Medium	High
Extra-org. contact	High	High	High	High	Low	High=>Low	High	High	High
Informal socialising	High	Medium	Medium	High	Low	High	High	Low	High
Mobile IT use	High	High	High	High	High	High	High	Medium	Medium
Visual surveillance	Medium	Low	Low	Low	Low	Low	Low	Low	High
Electronic surveillance	Medium	Medium	High	High	High	High	Medium	Medium	Low
Display status symbols	Low	Low	Medium	High	Low	Low	High	Low	High
Corporate aesthetic	High	High	High	High	Low	Medium	High	Medium	High

The focus on a single technology in everyday life, for example, implies less emphasis on the heterogeneous appropriation of service portfolio diversity, as the services that are predominantly used have traditionally been the standard ones offered by all mobile phone handsets. These have typically offered portable connectivity through voice calls, SMS messaging and mobile email. However, the popularity of smartphones with associated application and media-store platforms results in diverse usage patterns and greater innovation of mobile services diversity. With blurring boundaries between different spheres of life and the active role of mobile information technology across different spheres, it is critical to re-connect primarily technical discourses on the design of mobile information technology with the socio-technical understanding of the technology in action (Perlow, 1998; Golden and Geisler, 2007; Bassoli, 2010).

Furthermore, the emphasis on the use of mobile information technology in general social settings implies little or no emphasis on issues of collaboration and control, as the notion of mutual interdependencies and the management of activities are not core to the study of everyday activities. It is therefore natural that Ling's (2008) exploration of mobile telephony as a social activity emphasises rituals. Within the context of work, collaboration and control are essential aspects to consider in relation to technological support and issues widely discussed within a range of academic fields, such as IS, group decision support systems and CSCW. The challenge is here to understand the fundamental challenges of mobile information technology within the context of work.

Performances revisited

A range of paradoxes, contradictions, tensions and conflicting requirements governs organisational life. Organisations need to both explore new long-term options and exploit short-term gains. In order to do this, they must become contextually ambidextrous, which implies rendering some members able to balance the concerns within the context in which they find themselves.

This book considers the role of planned and emergent technology performances within such a context and explores the diversity of both mobile performances and related mobile service portfolios. Figure 8.1 illustrates the core tenet of the book; planned and emerging technology performances serve as the integrative meeting point of the contradictions of mobile work and mobile service portfolios. The mobile technology performances service purposes related to the overall cultivation of fluidity and boundaries. Such cultivation occurs across the activities related to individual creative interactional arrangements, the arrangement of horizontal team collaboration and control in mobile work.

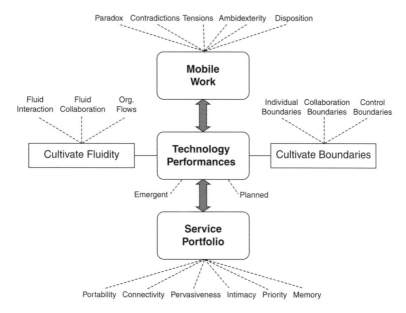

Figure 8.1 Overview of enterprise mobility

Creativity

At the level of individual creativity invested in the management of everyday interaction, this manifests itself as the cultivation of fluid interaction and of individual interaction barriers, by, for example:

- cultivating rhythms of interaction – for example, Ray the taxicab driver engaging and disengaging with his mobile phone depending on whether he has passengers on board;
- managing both the near and the remote as intimate connectivity allows remote requests to enter the situated interaction;
- converging separate interaction streams, which forms the key coping mechanism for Hiro the CEO;
- seeking to control unpredictable interaction – for example, by setting up filters;
- maintaining the balance of synchronous interaction by channelling it through asynchronous interaction, as Hiro's case also demonstrates;
- managing interruptions while remaining flexibly accessible, as both the cases of Hiro and the police officers John and Mary exemplify;
- seting up alarms triggered by external conditions, on which Khalid the trader spends considerable effort;

- remaining connected and reachable while mobile and anchored, which is the case whenever connected intimate and portable services are used;
- prioritising interaction through embedded asymmetry, as illustrated in a number of cases, perhaps most clearly in Hiro's case.

Cultivating fluid interaction implies the individual mobile worker seeking to optimise his or her own interaction so as to engage in fluid interaction. However, this can easily be challenged by emerging contingencies, unexpected interruptions and sudden needs to deviate from routines. This tension is more easily resolved for operational routine work, such as that of Simon the security guard, than for mobile work playing a key role in enabling contextual ambidexterity, such as in Hiro's case. Cultivating individual boundaries is a necessary part of cultivating fluid interaction in order to ensure availability whilst maintaining boundaries around this availability. The individual boundaries are highly individual yet still conflicting and are under constant construction informed by a wealth of influences, such as specific situations, moods and work pressures.

Collaboration

In terms of technology performances aimed at negotiating mutual interdependencies in collaboration, the tension translates into the cultivation of fluid collaboration and of collaborative boundaries through, for example:

- engaging in rhythms of collaboration, as illustrated by the shifting interactional intensity between John and Mary, the police officers, and remote parties;
- the distributed co-ordination of task allocation, as seen in the active queue displayed on the police vehicle MDT screen;
- enabling working alone while collaborating, best illustrated by the automated collaboration between Simon's security work and the central dispatch officer;
- instant access to central information, for example, provided by the MDT in the police vehicle;
- constant access to key colleagues, for example, provided through the shoulder-mounted personal police radio;
- ad hoc interaction to co-ordinate, exemplified by taxicab drivers calling colleagues to recruit cabs for emerging jobs;
- rendering remote activities transparent to facilitate mutual awareness in collaboration, as the security guard co-ordination mechanism illustrates;
- balancing narrow-casting to specific recipients and multi-casting to ensure mutual awareness, best illustrated by the balancing of a portfolio of mobile services used by the police officers;
- automating the ongoing relationship between remote collaborators and central reporting system, also illustrated by the security guard services;

- automatic resolution of mutual interdependencies to reduce the need for interaction, also illustrated by the security guard services;
- prioritising interaction between parties using embedded asymmetry, for example, implemented in Hiro's mobile phone filters;
- enabling real-time updates of remote status, something that is applied in both the RFID cases.

Cultivating fluid collaboration mirrors the individual aim to engage in fluid interaction and relates to the mobile worker's collaborative activities and articulation work in order to resolve mutual interdependencies. Tensions will easily emerge when differences in requirements, work styles and a range of other issues occur. Cultivating collaborative boundaries reflects the need for a current group of mutually interdependent mobile workers to ensure that their articulation of mutual interdependencies and collaborative activities can be conducted in a manner most conducive to ensuring the progress of their joint effort. This requires cultivation of the interactional boundary to outsiders. However, as collaborative arrangements are most often dynamic, the collaborative boundaries can constantly be up for negotiation. For example, in Simon's case (case 5), it was relatively stable, whereas for John and Mary (case 4), it was constantly shifting.

Control

For the technology performances related to the control of mobile work, the tension translates into the challenges of organising and cultivating mobile information flows and of cultivating organisational boundaries. For the mobile worker, this will, for example, imply:

- actively participating in cultivating mobile information flows, such as the police officers making decisions regarding when to use the radio and when to rely on the MDT;
- engaging in remotely planned performances while managing exceptions and retaining some local control, best illustrated by Jason's careful balancing of central requirements and emerging situations;
- documenting ephemeral interaction, exemplified by the requirement of Khalid to record trades on an answer machine;
- improvising performances when stuck between local and remote attempts to exercise control, illustrated by Yin's need to manage the conflict between local doctors and the remote learning centre;
- cultivating organisational boundaries through combinations of interaction symmetry and asymmetry, which, for example, is attempted in Jason's case.

The organisation of mobile information supply chains require balancing planned and emergent technology performances providing the contextually appropriate balance between the central and decentral control. In case 3, Khalid remained subjected to highly devolved control during out-of hours trading as the only feasible solution. For Winters in case 6 and Yin in case 9, the balancing of local and remote control was an issue of contention, whereas for Simon in case 5, strong central control was deemed most appropriate. Cultivating organisational boundaries is a key aspect of enterprise mobility, as mobile workers often represent edges of the organisation and therefore can help improve the contextual ambidexterity through balancing concerns for exploration and exploitation. In addition, for operational efficiency concerns, the careful cultivation of the organisation's interaction boundaries is both a challenging balance between facilitating planned and supporting emergent technology performances. Hiro's mobile phone in case 1 almost defines both the epicentre and boundary of his small organisation. By providing real-time data access to semi-automated status updates of Simon's progress in case 5, the client-facing services suddenly becomes much less work-intensive and much more interactively informative. For Jason in case 8, the operational issue of defining the organisational boundary is an issue of continual contention.

8.2   Managing mobility practices

Control through interaction

Yates (1989) argues that control can be obtained by socio-technical systems collecting and aggregating operational data on organisational performance upstream and communicating decisions and values downstream. Information technology serves as a means of satisfying the increasing need for greater sophistication of control as industrial societies became more complex and faster (Beniger, 1986). Enterprise mobility can be viewed as a response to further fragmentation and intensification of information work. Technological developments place powerful information and communication technologies directly in the hands of information workers. Organisations deploy these technologies in order to mobilise organisational capabilities through enabling access to corporate infrastructures and colleagues as well as to clients and customers. Most mobile services are not only symbolically but also physically associated with an individual user, following that user throughout his or her working day and in some cases outside of work as well. Shifts from functional hierarchies towards flatter modes of organising work leave more and more control out of the direct influence of bureaucratic procedures (Kunda, 1992, pp. 12ff). As the pace of work intensifies and as work is mobilised, the management of this work will increasingly rely on interactive processes of stipulating the outcome of work and of interacting in order to articulate the current state of affairs. Furthermore, the standardisation of skills and knowledge through professionalisation is a strong trend of the nineteenth and twentieth centuries (Perkin, 2002) and forms a common means for managing mobile work.

Mobile paradoxes

The challenges associated with enterprise mobility result in a number of paradoxes of mobile and remote working (Pearlson and Saunders, 2001; Dubé and Robey, 2009): increased flexibility aided by an increased structure; increased individualisation and collaboration; increased responsibility and control; the increased importance of the physical presence in mobile working; successful task-oriented collaboration through increased socialisation; and initial mistrust as a source of trust.

Organisations are paradoxically both able to increase the degree of organisational control and the degree of individual flexibility with mobile information technology. Enterprise mobility possibly implies both stronger and more intimate relationships forged between the user and the technology, as well as the risk of increasing alienation in cases of relentless requirements to interact and engage. In terms of the distinction between internal and external concerns (Quinn and Rohrbaugh, 1983), organisations paradoxically obtain a greater ability to cultivate intelligent organisational boundaries and encounter increased difficulties in actually doing so. Enterprise mobility both facilitates placing centralised corporate infrastructures and services directly in the hands of organisational members as well as allowing these members to foster new emerging relationships beyond those that are centrally controlled. This relates directly to the possibilities of rendering participants and, in turn, the organisation contextually ambidextrous.

Resolving such paradoxes in specific contexts requires a combination of one or more of these four strategies (Poole and Van de Ven, 1989; Beech *et al.*, 2004): *transform* – accepting the paradox and learning to live with it; *eliminate* – resolving the paradox through spatial separation; *avoid* – resolving the paradox by situating at different temporal locations; and *transcend* – resolving the paradox through synthesis. As an example, accepting the paradox of control versus flexibility can result in establishing planned improvisation, which implies setting boundaries but assuming that they can only be contextually resolved (Pearlson and Saunders, 2001, p. 121). Similarly, avoiding the same paradox can be resolved by ensuring that participants plan their availability for when they can interact.

Mobile performance requirements and responses

In the meeting of organisational paradoxes, contradictions and conflicting demands, the individual mobile worker will engage in technology performances to support communication, decisions, collaboration and the articulation of work to resolve mutual interdependencies. Mintzberg (1983, p. 108) argues that there is a continuum of co-ordination from the horizontally centralised use of direct supervision, through the standardisation of work processes, the standardisation of outputs and the standardisation of skills, to the horizontally decentralised form of mutual adjustment. This is similar to the discussion regarding the degree of local control in the co-ordination of mutual interdependent activities spanning a high degree of local control through informal discussion to a low degree of local control enforced through standardised and stipulated co-ordination mechanisms (Schmidt, 1993, p. 91).

Local control – discretion – in the response to emergence does not necessarily match the need for emerging decisions with emerging technology performances resolving the situation, as illustrated in Figure 8.2. The requirement to improvise may indeed be met with a responsive performance, but can equally result in an unresponsive performance. The organisational expectation to participate in a planned performance can result in an improvised performance that may or indeed may not be desirable for the organisation.

Individualisation of work

The past decades have seen the changing granularity of information work, first resulting from office and factory automation in the 1970s and 1980s, the shift towards teams from the 1980s and an individualisation of work activities from the 1990s (Zuboff, 1988; Randall *et al.*, 2007). Enterprise mobility contributes to further this process by increasing individualisation of work yet increasing opportunities for tighter interactive coupling between technology performances and underlying organisational/team control systems. This does not

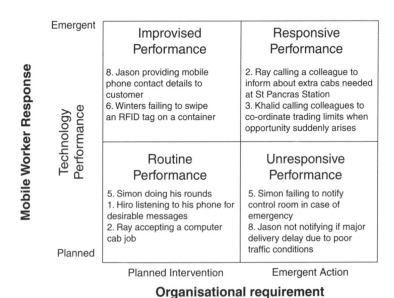

Figure 8.2 Diversity of enterprise mobility performances in terms of planned interventions and emergent actions

necessarily imply more control; rather, that emergence can be recognised, recorded and acted upon more easily. Enterprise mobility contributes significantly to Beniger's (1986) control revolution.

The extensive individualisation and intimacy in technological support for information management and organisational communication co-evolves with increased individualisation of work and its co-ordination and control. Contemporary developments, for example within the software industry, engage processes of output standardisation allowing for much more individual and independent activities to be assembled into a collaborative effort without extensive ongoing informal mutual adjustments (Voutsina *et al.*, 2007; Voutsina, 2008).

Myriads of intricate connections and aggregations of distributed data into management information replace or augment central computational models, making the control system far more adaptive to locally emerging needs to nudge the stipulated constraints and procedures slightly. However, systems can still contain significant centralisation – for example, by goals imposed from the centre (Courpasson, 2000).

The expectation of being subjected to some form of continuous monitoring and appraisal naturally follows traditional office or workstation working, and the absence of this if mobile working also implies increased individualisation of work can result in workers feeling isolated. Whittle (2005, p. 1314) reports a member of a remotely distributed team stating that he could be dead for two weeks without his boss ever knowing it.

Mobile ambidexterity

The management and resolution of organisational paradoxes tensions through enterprise mobility can relate to ensuring operational efficiency but can also serve the purpose of enabling contextual ambidexterity through supporting individuals to: take the initiative and identify opportunities; cooperate and seek to develop joint opportunities; broker and build relationships; and multitask (Birkinshaw and Gibson, 2004, p. 49). Birkinshaw and Gibson (2004, p. 51) suggest the importance of providing social support and managing performance. Balancing these concerns relates directly to balancing formal and personal support for decision-making (Mom *et al.*, 2009), which again relates directly to the debate of the role of formal structures in organisations (Schmidt, 1999).

8.3 Conclusion

This book has discussed the application of a diversity of mobile information technology in the form of mobile services at work. Nine case studies of enterprise mobility – mobile services at work – were discussed with respect to three distinct concerns: 1) everyday creativity in the organisation of interaction;

2) collaborative interaction related to the negotiation of mutual interdependencies; and 3) interaction serving the purpose of controlling and managing mobile work activities.

Coping with paradoxes by cultivating fluidity and boundaries

The mobile information worker copes in his or her daily work with tensions, paradoxes, contradictions and conflicting demands. Dealing with these requires the application of a mixture of planned interventions and emergent actions. For the individual mobile information worker, this requires support for balancing planned and emergent technology performances so as to counter these conflicting demands. Enterprise mobility portfolios denote the diversity of mobile services representing contradicting and redundant affordances, constraints and services to be contextualised and evoked by the mobile information worker.

Mobile service portfolios offer selective combinations of services comprising affordances, constraints and mechanisms. Affordances denote the analytical characterisation of services that provide resources for reflexive action by the user, as opposed to mechanisms that are scripts guiding or stipulating action. The distinction is analytical as in some situations the user can treat a mechanism as an affordance as well as define an affordance into a mechanism. Service portfolios combine the opportunities for: *intimacy* – close bodily connection with services and the identification of the user; *connectivity* between the user and remote users and services; *priority* of interaction; *pervasiveness*, with services relying on environmental information; *memory* providing support for ongoing relationships; and *portability* – the ability to carry the service along.

Planned and emergent technology performances denote contrary, not linear, relationships between service and purpose. The mobile worker will in any given situation be subjected to a range of opportunities, constraints, demands and desires. The instantiation of specific service combinations with specific data configurations will depend on the chosen resolution of planned and emerging technology performances.

Planned and emerging technology performances

The primary paradox in enterprise mobility performances is between responding to the conflicting requirements of planned interventions and emerging actions by both cultivating fluidity and boundaries. For individual interaction management, this translates into the creative cultivation of both fluid interaction and individual boundaries. For the negotiation of mutual interdependencies in collaboration, the mobile worker will participate in the cultivation of fluid collaboration and collaborative boundaries. In terms of the management and control of mobile working, the paradox requires both the orchestration of mobile interaction flows and the cultivation of organisational boundaries.

For mobile information workers with the additional requirement of not only coping with conflicting operational requirement but also actively enabling the organisation to be contextually ambidextrous, enterprise mobility can provide key support through services making it possible to balance emergent and planned technology performances, allowing the resolution of emerging tensions and contradictions.

Research challenges

Several research traditions tackle a variety of aspects within the broader arena of mobilities research. To what extent is cross-fertilisation between these research strands possible or even desirable? The 'mobility turn' (Urry, 2007) can be interpreted as a conceptual clearing acting as a timely response to a range of global phenomena. As a conceptual clearing, it is forgiving in allowing a diverse, heterogeneous and sprawling research debate. Broad and open-ended research discussions centred on a uniting but ill-defined conceptual clearing can potentially bring together researchers and practitioners that under normal circumstances, given a well-structured and well-defined research agenda, would not engage in a common debate. The conceptual clearing will allow a different combination of skills, facilitate access to alternative research funding pools, help the questioning of assumptions and possibly even make conferences more interesting to attend.

Over time it is likely that the broad and sprawling academic discourse defined by a conceptual clearing will be the subject of intense discussions and as a result will deliver some consensus within the community of the definition of key terms, core commonly held assumptions, research approaches that are considered appropriate, and a prioritisation of core issues to be researched. It is also likely that the more mature such debate becomes, the more parts of the community may decide that they do not see themselves as contributing further to the field, with the resulting fragmentation of the community. In terms of understanding mobility and information technology within the context of work, it can be argued that there is a need for a broader community to be formed.

One example of cross-fertilisation from the general mobilities discussion across to enterprise mobility research is the expansion of the mobility concept beyond the focus on corporeal movement and instead emphasising the mobilisation of interaction at work through a variety of technologies (Kakihara and Sørensen, 2002). This broadening of the analytical scope brought with it an interesting understanding of the roles of technology and mobility, for example, emphasising the role of mobile communications in terms of what Ling (2004) calls micro-co-ordination and augmenting and arranging essential situated interaction as well as a connected presence (Licoppe, 2004) amongst remote colleagues. Within much of the technology-centred discussion of mobility there still is a strong paradigm of one-sided emphasis on technology affordances, broadly translated into the mantra of 'anytime, anywhere, anyone' interactivity. This is clearly an assumption in dire need of anchoring in a set of alternative assumptions related to situated human practices and emotions (Ciborra, 2006).

However, it can equally be argued that the separate academic communities each discussing their own aspect mobilities should rather seek to further narrow and deepen their discourse rather than broaden it. Each tradition is possibly associated with several (conflicting) sets of assumptions and the black-boxing of some aspects in order to highlight others – for example, sociology black-boxing mobilities (Urry, 2007), management research black-boxing materiality (Orlikowski and Scott, 2008), CSCW research black-boxing mobility,[17] mobile phone studies neglecting work (Sørensen, 2011) and so on.

Enterprise mobility as the academic study of mobile information technology at work is yet to transform into a conceptual clearing. While most organisations employ workers who use mobile information technology and a growing proportion of the working population engages in mobile information work, the IS field is only slowly catching up and recognising the need to address these issues. This book has been an attempt to help kick-start such discussion. Given the spiralling development of new working practices using new and more advanced mobile service portfolios, there is no end in sight for the practitioner of innovations fuelling the research debate. Let us hope that research follows suit and both helps to make sense of developments and informs future paths to be taken.

Notes

1. 2010 figures. Updated statistics from the GSM Association are available on www. gsmamobileinfolink.com. Please note that connections and handsets are not the same as there are less than this number of active handsets but much more handsets produced in total.
2. *Management Information Systems Quarterly (MISQ), Information Systems Research (ISR), Journal of Management Information Systems (JMIS), Journal of the Association of Information Systems (JAIS), Journal of Information Technology (JIT), Journal of Strategic Information Systems (JSIS), Information Systems Journal (ISJ)* and *European Journal of Information Systems (EJIS)* in order of impact factor. This list of journals has been selected by senior IS scholars and more information can be found here: http://home.aisnet.org/displaycommon. cfm?an=1&subarticlenbr=346 (date accessed 26 April 2011).
3. The basket of eight IS journals does of course not signify all research on mobile information technology in general and IS research in particular. A quick search on all volumes of some of the remaining IS journals for titles or abstracts containing either the term 'mobile' or 'cell', plus adding articles found manually, identified the following number of articles on mobile information technology: *Communications of the Association of Information Systems* (one), *Behaviour and Information Technology* (two), *Information Technology & People* (two), *Scandinavian Journal of Information Systems* (two) and *Information and Organization and Journal of Computer Supported Cooperative Work* (eight). The journals dedicated to mobile and ubiquitous computing, such as *IEEE Pervasive Computing, Personal and Ubiquitous Computing* and the *Journal of Mobile Marketing*, as well as the HCI journals, have many more articles for obvious reasons. However, the discourse in these journals will also often be remote from the core concerns within Management Information Systems (MIS), as for example illustrated by Sørensen and Al-Taitoon (2008), extending the usability concept to encompass organisational issues.
4. See Malone (2004) for interesting calculations of the differences in transaction costs of sending one page of text by pre-railroad mail, by rail, by telegraph or by email. For 100 destinations the cost in the 1840s was $100 by pre-railroad mail and $750 by telegraph in the 1850s.
5. Gibson and Sterling (1991) explore what the world might have looked like if engineering tolerances in Baggage's time would have allowed for the construction of his mechanical computers. They paint a picture of extensive London smog as steam engines produce enough energy to operate large amount of mechanical computers: http://en.wikipedia.org/wiki/The_Difference_Engine (date accessed 26 April 2011).
6. http://en.wikipedia.org/wiki/Microsoft (date accessed 27 April 2011).
7. The photo was distributed from Twitter users to blogs and news sites within minutes of being uploaded: www.twitpic.com/135xa (date accessed 27 April 2011).
8. The distinction between affordances and mechanisms is inspired by the discussion with Christian Licoppe over a nice lunch one spring day in Paris, 2010, after I presented my work to his department.
9. Other researchers have characterised their challenges similarly – for example, Aaen's (1989) discussion of systems development itself and Gherardi's (2006) characterisation of 'knowing in practice' as the balancing of a mentalistic vision of knowledge against its commodification.

10. As an example, the CSCW community journal has since its inception in 1991 published around a dozen articles on issues directly relating to mobile information technology and/or mobile working, giving rise to a call from within the community to move with the technological times (Crabtree *et al.*, 2005). In April 2008 the *CSCW Journal* published a special issue on the role of place in collaboration (Ciolfi *et al.*, 2008).
11. A simple Google Scholar search for 'anytime, anywhere' on 29 April 2010 resulted in 31,900 references across a range of fields and disciplines, but mostly related to communication and interaction through Internet and mobile technologies.
12. A Google search for 'crackberry' on 29 April 2010 yielded over one million pages, whilst searching Google Scholar using the same term resulted in 297 pages.
13. The following description will assume a clear division of responsibilities between the victim and perpetrator for the sake of simplicity, as any use of the two terms 'victim' or 'perpetrator' would otherwise have to be prefixed with 'alleged'.
14. Services such as foursquare.com and gowalla.com provide proximity- or location-based social networking.
15. The saturated graph is n*(n-1)/2, with n being number of nodes. Six service categories imply 15 category pairs. With each category denoting a basic binary choice, the saturated graph with 12 nodes contains 66 pairs, of which the six inter-category connections do not make sense – for example, encounter <-> relationship. This implies a total of 60 possible service element pairs.
16. In this respect, the portfolio can be seen as the class from which the object is instantiated and populated with data through interaction.
17. Based on the relatively small number of papers published that deal directly with issues related to the support for mobile working.

References

Aaen, I. (1989) 'Systemudvikling: Mellem Skylla og Charybdis' ('Systems Development: Between Skylla and Charybdis'). PhD dissertation, Aalborg University.

Abowd, G.D. and Mynatt, E.D. (2000) 'Charting Past, Present, and Future Research in Ubiquitous Computing', *ACM Transactions on Computer-Human Interaction*, 7(1): 29–58.

Ackroyd, S. (1992) *New Technology and Practical Police Work: The Social Context of Technical Innovation*. Buckingham: Open University Press.

Adomavicius, G., Bockstedt, J., Gupta, A. and Kauffman, R. (2007) 'Technology Roles and Paths of Influence in an Ecosystem Model of Technology Evolution', *Information Technology and Management*, 8(2): 185–202.

——. (2008) 'Making Sense of Technology Trends in the Information Technology Landscape: A Design Science Approach', *MIS Quarterly*, 32(4): 779–809.

Aducci, R., Bilderbeek, P., Brown, H., Dowling, S., Freedman, N., Gantz, J., Germanow, A., Manabe, T., Manfrediz, A. and Verma, S. (2008) 'The Hyperconnected: Here They Come! A Global Look at the Exploding "Culture of Connectivity" and its Impact on the Enterprise', IDC White Paper sponsored by Nortel.

Agar, J. (2003) *Constant Touch: A Global History of the Mobile Phone*. Cambridge: Icon Books.

Ahuja, M.K., McKnight, D.H., Chudoba, K.M., George, J.F. and Kacmar, C.J. (2007) 'IT Road Warriors: Balancing Work–Family Conflict, Job Autonomy, and Work Overload to Mitigate Turnover Intentions', *MIS Quarterly*, 31(1): 1–17.

Al-Taitoon, A. (2005) 'Making Sense of Mobile ICT-Enabled Trading in Fast Moving Financial Markets as Volatility-Control Ambivalence: Case Study on the Organisation of Off-Premises Foreign Exchange at a Middle-East Bank', PhD dissertation, London School of Economics.

Al-Taitoon, A. and Sørensen, C. (2004) 'Supporting Mobile Professionals in Global Banking: The Role of Global ICT-Support Call-Centres', *Journal of Computing and Information Technology*, 12(4): 297–308.

Albrecht, K. and McIntyre, L. (2006) *Spychips: How Major Corporations and Government Plan to Track Your Every Move with RFID*. USA: Nelson Current.

Andersen, P.B., Holmqvist, B. and Jensen, J.F. (eds) (1993) *The Computer as Medium*. Cambridge University Press.

Andersson, P., Essler, U. and Thorngren, B. (eds) (2007) *Beyond Mobility*. Malmo, Sweden: Studentlitteratur.

Andriessen, J.H.E. and Vartiainen, M. (eds) (2005) *Mobile Virtual Work: A New Paradigm?* Berlin: Springer.

Andriopoulos, C. and Lewis, M.W. (2010) 'Managing Innovation Paradoxes: Ambidexterity Lessons from Leading Product Design Companies', *Long Range Planning*, 43(1): 104–22.

Angus, A., Papadogkonas, D., Papamarkos, G., Roussos, G., Lane, G., Martin, K., West, N., Thelwall, S., Sujon, Z. and Silverstone, R. (2008) 'Urban Social Tapestries', *IEEE Pervasive Computing*, 7(4): 44–51.

Anonymous (1870) 'The Police of London', *Quarterly Review (London)*, 129: 90–129.

Antero, M. (2006) 'MICTs in Technology Consulting: An Assessment of their Level of Ubiquity, Impact and Usage Among the Industry's Knowledge Workers', MSc dissertation, London School of Economics.

Aoki, M. and Dore, R. (1996) *The Japanese Firm: The Sources of Competitive Strength*. Oxford University Press.

Arnold, M. (2003) 'On the Phenomenology of Technology: The "Janus-Faces" of Mobile Phones', *Information and Organization*, 13: 231–56.

Arnold, M., Shepherd, C. and Gibbs, M. (2008) 'Remembering Things', *Information Society*, 24(1): 47–53.

Arora, R. (2003) 'Professional Salesmen Utilization of Mobile Technology: An Empirical Case Study', MSc dissertation, Department of Management, Information Systems and Innovation Group, London School of Economics.

Ashforth, B.E., Kreiner, G.E. and Fugate, M. (2000) 'All in a Day's Work: Boundaries and Micro Role Transition', *Academy of Management Review*, 25(3): 472–91.

Avgerou, C., Ciborra, C. and Land, F. (ed.) (2004) *The Social Study of Information and Communication Technology: Innovation, Actors and Contexts*. Oxford University Press.

Axtell, C., Hislop, D. and Whittaker, S. (2008) 'Mobile Technologies in Mobile Spaces: Findings from the Context of Train Travel', *International Journal of Human-Computer Studies*, 66(12): 902–15.

Baecker, R.M. (ed.) (1993) *Readings in Groupware and Computer-Supported Cooperative Work Assisting Human-Human Collaboration*. San Francisco: Morgan Kaufmann.

Baldwin, C.Y. and Clark, K.B. (2000) *Design Rules: The Power of Modularity, Volume 1*. Cambridge, MA: MIT Press.

Baresi, L., Dustdar, S., Gall, H. and Matera, M. (eds) (2004) *Ubiquitous Mobile Information and Collaboration Systems: Second CAiSE Workshop, UMICS 2004, Riga, Latvia*. Berlin: Springer-Verlag.

Barfield, W. and Caudell, T. (eds) (2001) *Fundamentals of Wearable Computers and Augmented Reality*. Hillsdale, NJ: Lawrence Erlbaum Associates.

Barley, S.R. (1988) 'On Technology, Time and Social Order: Technically Induced Change in the Temporal Organization of Radiological Work', in F.A. Dubinskas (ed.), *Making Time: Ethnographies of High-Technology Organizations*. Philadelphia: Temple University Press, pp. 123–69.

Barley, S.R. and Kunda, G. (2004) *Gurus, Hired Guns, and Warm Bodies: Itinerant Experts in a Knowledge Economy*. Princeton University Press.

Barnes, S. (2003) 'Enterprise Mobility: Concept and Examples', *International Journal of Mobile Communications*, 1(4): 341–59.

Baron, N.S. (2008) *Always On: Language in an Online and Mobile World*. Oxford University Press.

Barrett, M.I. and Scott, S.V. (2004) 'Electronic Trading and the Process of Globalization in Traditional Futures Exchanges: A Temporal Perspective', *European Journal of Information Systems*, 13(1): 65–79.

Basole, R.C. (ed.) (2008) *Enterprise Mobility: Applications, Technologies and Strategies*. Special issue on Enterprise Mobility of the Information Knowledge Systems Management Journal, IOS Press.

——. (2009) 'Visualization of Interfirm Relations in a Converging Mobile Ecosystem', *Journal of Information Technology*, 24: 144–59.

Bassoli, A. (2010) 'Living the Urban Experience: Implications for the Design of Everyday Computational Technologies', PhD dissertation, London School of Economics and Political Science.

Bassoli, A., Brewer, J., Dourish, P., Martin, K. and Mainwaring, S. (2007) 'Underground Aesthetics: Rethinking Urban Computing', *IEEE Pervasive Computing*, 6(3): 39–45.

Batt, R. and Doellgast, V. (2005) 'Groups, Teams, and the Division of Labor', in S. Ackroyd, R. Batt, P. Thompson and P.S. Tolbert (eds), *The Oxford Handbook of Work and Organization*. Oxford University Press, pp. 138–61.

Bauman, Z. (2000) *Liquid Modernity*. Cambridge: Polity Press.

——. (2007) *Liquid Times*. Cambridge: Polity Press.

BBC News (2003) 'The Cause of the BA Dispute', http://news.bbc.co.uk/1/hi/business/3093099.stm, date accessed 6 May 2011.

Becker, F. (2004) *Offices at Work: Uncommon Workspace Strategies that Add Value and Improve Performance*. New York: Jossey-Bass Business & Management.

Beech, N., Burns, H., de Caestecker, L., MacIntosh, R. and MacLean, D. (2004) 'Paradox as Invitation to Act in Problematic Change Situations', *Human Relations*, 57(10): 1313–32.

Bell, D. (1976) *The Coming of Post-Industrial Society: A Venture in Social Forecasting*. Harmondsworth: Penguin.

Beniger, J.R. (1986) *The Control Revolution: Technological and Economic Origins of the Information Society*. Harvard University Press.

Berghel, H. (1999) 'Digital Village: The Cost of Having Analog Executives in a Digital World', *Communications of the ACM*, 42(11): 11–15.

Birkinshaw, J. and Gibson, C. (2004) 'Building Ambidexterity into an Organization', *Sloan Management Review*, 45(4): 47–55.

Bittman, M., Brown, J.E. and Wajcman, J. (2009) 'The Mobile Phone, Perpetual Contact and Time Pressure', *Work, Employment & Society*, 23(4): 673–91.

Bittner, E. and Bish, R.L. (1975) *The Functions of the Police in Modern Society: A Review of Background Factors, Current Practices, and Possible Role Models*. Seattle: University of Washington Press.

Blackler, F. (1995) 'Knowledge, Knowledge Work and Organizations: An Overview and Interpretation', *Organization Studies*, 16: 1021–46.

Boateng, K. (2010) 'ICT-Driven Interactions: On the Dynamics of Mediated Control', PhD dissertation, London School of Economics.

Bolter, J.D. (1982) *Turing's Man: Western Culture in the Computer Age*. Chapel Hill, NC: University of North Carolina Press.

Broadbent, M., Weill, P. and St. Clair, D. (1999) 'The Implications of Information Technology Infrastructure for Business Process Redesign', *MIS Quarterly*, 23(2): 159–82.

Brodt, T.L. and Verburg, R.M. (2007) 'Managing Mobile Work: Insights from European Practice', *New Technology, Work and Employment*, 22(1): 52–65.

Brown, B. and O'Hara, K. (2003) 'Place as a Practical Concern of Mobile Workers', *Environment and Planning A*, 35(9): 1565–88.

Bullen, C. and Bennett, J. (1990) 'Groupware in Practice: An Interpretation of Work Experience', MIT Center for Information Systems Research, Cambridge, MA, March.

Bunting, M. (2004) *Willing Slaves: How the Overwork Culture is Ruling Our Lives*. London: HarperCollins.

Bunzel, D. (2002) 'The Rythm of the Organization: Simultaneity, Identity and Discipline in an Australian Coastal Hotel', in R. Whipp, B. Adam and I. Sabelis (eds), *Making Time: Time and Management in Modern Organizations*. Oxford University Press, pp. 168–81.

Burton-Jones, A. and Gallivan, M.J. (2007) 'Toward a Deeper Understanding of System Usage in Organizations: A Multilevel Perspective', *MIS Quarterly*, 31(4): 657–79.

Büscher, M., Urry, J. and Witchger, K. (eds) (2011) *Mobile Methods*. Abingdon: Routledge.

Cairncross, F. (1997) *The Death of Distance: How the Communications Revolution will Change our Lives*. Boston, MA: Harvard Business School Press.

Cameron, K.S. and Quinn, R.E. (1988) *Paradox and Transformation: Toward a Theory of Change in Organization and Management*. New York: Harper Business.

Caminer, D., Aris, J., Hermon, P. and Land, F. (1998) *L.E.O. – The Incredible Story of the World's First Business Computer*. London: McGraw-Hill Education.

Carlstein, T. (1983) *Time Resources, Society and Ecology*. London: George Allen & Unwin.

Carnoy, M. (2002) *Sustaining the New Economy: Work, Family and Community in the Information Age*. Harvard University Press.

Carr, N.G. (2009) 'The Eternal Conference Call', http://www.roughtype.com/archives/2009/10/the_eternal_con.php, date accessed 6 May 2011.

——. (2010) *The Shallows: What the Internet is Doing to our Brains*. London: W.W. Norton & Co.

Carstensen, P. and Sørensen, C. (1996) 'From the Social to the Systematic: Mechanisms Supporting Coordination in Design', *Journal of Computer Supported Cooperative Work*, 5(4): 387–413.

Casey, E.S. (2009) *Getting Back into Place: Toward a Renewed Understanding of the Place-World*. Bloomington, IN: Indiana University Press.

Castells, M. (1996) *The Rise of the Network Society*, Vol. 1. The Information Age: Economy, Society and Culture. Oxford: Blackwell.

Castells, M., Qiu, J.L., Fernandez-Ardevol, M. and Sey, A. (2007) *Mobile Communication and Society: A Global Perspective*. Cambridge, MA: MIT Press.

Cha, H.S., Pingry, D.E. and Thatcher, M.E. (2008) 'Managing the Knowledge Supply Chain: An Organizational Learning Model of Information Technology Offshore Outsourcing', *MIS Quarterly*, 32(2): 281–306.

Chae, B. and Bloodgood, J. (2006) 'The Paradoxes of Knowledge Management: An Eastern Philosophical Perspective', *Information and Organization*, 16(1): 1–26.

Chemero, A. (2003) 'An Outline of a Theory of Affordances', *Ecological Psychology*, 15(2): 181–95.

Chipchase, J., Aarras, M., Persson, P., Yamamoto, T. and Piippo, P. (2004) 'Mobile Essentials: Field Study and Concepting', in *Proceedings of the 2005 Conference on Designing for User eXperience (DUX '05)*. New York: AIGA (American Institute of Graphic Arts), Article 57.

Choi, T.Y., Budny, J. and Wank, N. (2004) 'Intellectual Property Management: A Knowledge Supply Chain Perspective', *Business Horizons*, 47(1): 37–44.

Ciborra, C. (1993) *Teams, Markets and Systems. Business Innovation and Information Technology*. Cambridge University Press.

——. (ed.) (1996) *Groupware and Teamwork*. Chichester: John Wiley & Sons.

——. (1997) 'De Profundis? Deconstructing the Concept of Strategic Alignment', *Scandinavian Journal of Information Systems*, 9(1): 67–82.

——. (2002) *The Labyrinths of Information: Challenging the Wisdom of Systems*. Oxford University Press.

——. (2006) 'The Mind or the Heart? It Depends on the (Definition of) Situation', *Journal of Information Technology*, 21: 129–39.

Ciborra, C. and Associates (2000) *From Control to Drift: The Dynamics of Corporate Information Infrastructures*. Oxford University Press.

Ciborra, C. and Patriotta, G. (1996) 'Groupware and Teamwork in New Product Development: The Case of a Consumer Goods Multinational', in C. Ciborra (ed.), *Groupware and Teamwork*. Chichester: John Wiley & Sons, pp. 121–44.

Ciolfi, L., Fitzpatrick, G. and Bannon, L. (2008) 'Settings for Collaboration: The Role of Place', *Computer Supported Cooperative Work (CSCW)*, 17(2): 91–6.

Clark, M. (2002) *Paradoxes from A to Z*. London: Routledge.

Clegg, S., Courpasson, D. and Phillips, N. (2006) *Power and Organizations*. London: Sage.

Close, F. (2001) *Lucifer's Legacy: The Meaning of Asymmetry*. Oxford: Paperbacks.

Coase, R.H. (1937) 'The Nature of the Firm', *Economica N.S.*, 4(4): 386–405.

Collins, R. (2004) *Interaction Ritual Chains*. Princeton University Press.

Conradson, B. (1988) *Kontorsfolk: Etnologiska bilder af livet på kontoret (Office People: Ethnographic Images of Life at the Office)*. Stockholm: Nordiska Muséet.

Cook, S.D.N. and Brown, J.S. (1999) 'Bridging Epistemologies: The Generative Dance Between Organizational Knowledge and Organizational Knowing', *Organization Science*, 10(4): 381–400.

Courpasson, D. (2000) 'Managerial Strategies of Domination. Power in Soft Bureaucracies', *Organization Studies*, 21(1): 141–61.

Cousins, K.C. (2004) 'Access Anytime, Anyplace: An Empirical Investigation of Patterns of Technology Use within Nomadic Computing Environments', PhD dissertation, Mack Robinson College of Business, Georgia State University, Atlanta.

Cousins, K.C. and Robey, D. (2005) 'Human Agency in a Wireless World: Patterns of Technology Use in Nomadic Computing Environments', *Information and Organization*, 15: 151–80.

Cousins, K.C. and Varshney, U. (2009) 'Designing Ubiquitous Computing Environments to Support Work Life Balance', *Communications of the ACM*, 52(5): 117–23.

Coviello, N.E. and Brodie, R.J. (2001) 'Contemporary Marketing Practices of Consumer and Business-to-Business Firms: How Different are They?', *Journal of Business and Industrial Marketing*, 16(5): 382–400.

Crabtree, A., Rodden, T. and Benford, S. (2005) 'Moving with the Times: IT Research and the Boundaries of CSCW', *Computer Supported Cooperative Work (CSCW)*, 14(3): 217–51.

Cresswell, T. (2006) *On the Move: Mobility in the Modern Western World*. New York: Routledge.

Csikszentmihalyi, M. (2002) *Flow: The Classic Work on How to Achieve Happiness*. London: Rider & Co.

Daft, R.L. and Lengel, R.H. (1986) 'Organizational Information Requirements, Media Richness and Structural Design', *Management Science*, 32(5): 554–71.

Dale, K. and Burrell, G. (2008) *The Spaces of Organisation and the Organisation of Space: Power, Identity and Materiality at Work*. Basingstoke: Palgrave Macmillan.

Daniels, K., Lamond, D. and Standen, P. (2001) 'Teleworking: Frameworks for Organizational Research', *Journal of Management Studies*, 38(8): 1151–86.

Darden, J.A. (ed.) (2009) *Establishing Enterprise Mobility at Your Company*. New York: Thomson Reuters/Aspatore.

Davis, G.B. (2002) 'Anytime/Anyplace Computing and the Future of Knowledge Work', *Communications of the ACM*, 45(12): 67–73.

Desbarats, G. (1999) 'The Innovation Supply Chain', *Supply Chain Management: An International Journal*, 4(1): 7–10.

Dijkstra, E. (1968) 'Go To Statement Considered Harmful', *Communications of the ACM*, 11(3): 147–8.

Dindia, K. (1987) 'The Effects of Sex of Subject and Sex of Partner on Interruptions', *Human Communication Research*, 13(3): 345–71.

Dix, A., Rodden, T., Davies, N., Trevor, J., Friday, A. and Palfreyman, K. (2000) 'Exploiting Space and Location as a Design Framework for Interactive Mobile Systems', *ACM Transactions on Computer-Human Interaction*, 7(3): 285–321.

Donner, J. (2008) 'Research Approaches to Mobile Use in the Developing World: A Review of the Literature', *Information Society*, 24(3): 140–59.

Dourish, P. (2001) *Where the Action Is: The Foundations of Embodied Interaction*. Cambridge, MA: MIT Press.

Dourish, P. and Bell, G. (2007) 'The Infrastructure of Experience and the Experience of Infrastructure: Meaning and Structure in Everyday Encounters with Space', *Environment and Planning B: Planning and Design*, 34(3): 414–30.

Dreyfus, H.L., Dreyfus, S.E. and Athanasiou, T. (1986) *Mind over Machine: The Power of Human Intuition and Expertise in the Era of the Computer*. Oxford: Basil Blackwell.

Dubé, L. and Robey, D. (2009) 'Surviving the Paradoxes of Virtual Teamwork', *Information Systems Journal*, 19(1): 3–30.

Eaton, B., Elaluf-Calderwood, S. and Sørensen, C. (2010) 'A Methodology for Analysing Business Model Dynamics for Mobile Services using Control Points and Triggers', in S. Sharrock and P. Ballon (eds), *Business Models for Mobile Platforms (BMMP 2010)*, Berlin: IEEE.

The Economist (2005) 'The CrackBerry Backlash', 25 May.

——. (2007) *Pocket World in Figures: 2007 Edition*. London: The Economist.

——. (2009) 'Mobile Marvels: A Special Report on Telecoms in Emerging Markets', 26 September.

——. (2010) 'Data, Data Everywhere: A Special Report on Managing Information', 27 February.

Edström, A. and Galbraith, J. (1977) 'Transfer of Managers as a Coordination and Control Strategy in Multinational Organizations', *Administrative Science Quarterly*, 22(2): 248–63.

Ehn, P. (1988) *Work-Oriented Design of Computer Artifacts*. Stockholm: Arbetslivscentrum.

Ehn, P. and Kyng, M. (1985) *A Tool Perspective on Design of Interactive Computer Support for Skilled Workers*. DAIMI-Århus University: Computer Science Department.

Eisenhardt, K.M. (1989) 'Making Fast Strategic Decisions in High-Velocity Environments', *Academy of Management Journal*, 32: 543–76.

Elaluf-Calderwood, S. (2008) 'Organizing Self-Referential Taxi –Work with mICT: The Case of the London Black Cab Drivers', PhD dissertation, London School of Economics.

Elaluf-Calderwood, S., Kietzmann, J. and Saccol, A. Z. (2005) 'Methodological Approach for Mobile Studies: Empirical Research Considerations', *4th European Conference on Research Methods in Business and Management*, Université Paris-Dauphine, Paris, France, 21–22 April.

Elaluf-Calderwood, S. and Sørensen, C. (2008) '420 Years of Mobility: ICT Enabled Mobile Interdependencies in London Hackney Cab Work', in D. Hislop (ed.), *Mobility and Technology in the Workplace*. London: Routledge, pp. 135–50.

Ellis, C.A., Gibbs, S.J. and Rein, G.L. (1991) 'Groupware: Some Issues and Experiences', *Communications of the ACM*, 34(1): 38–58.

Engeström, Y. (2008) *From Teams to Knots: Studies of Collaboration and Learning at Work*. Cambridge University Press.

Eppler, M.J. and Mengis, J. (2004) 'The Concept of Information Overload: A Review of Literature from Organization Science, Accounting, Marketing, MIS and Related Disciplines', *Information Society*, 20: 325–44.

Ericson, R.V. and Haggerty, K.D. (1997) *Policing the Risk Society*. University of Toronto Press.

Ericsson, K.A., Krampe, R.T. and Tesch-Römer, C. (1993) 'The Role of Deliberate Practice in the Acquisition of Expert Performance', *Psychological Reviews*, 100: 363–406.

Farson, R. (1996) *Management of The Absurd: Paradoxes in Leadership*. New York: Simon & Schuster.

Felstead, A. and Jewson, N. (2000) *In Work, at Home: Towards an Understanding of Homeworking*. London: Routledge.

Felstead, A., Jewson, N. and Walters, S. (2005) *Changing Places of Work*. London: Palgrave Macmillan.

Ferneley, E. and Light, B. (2008) 'Unpacking User Relations in an Emerging Ubiquitous Computing Environment: Introducing the Bystander', *Journal of Information Technology*, 23: 163–75.

Floridi, L. (2007) 'A Look into the Future Impact of ICT on our Lives', *Information Society*, 23(1): 59–64.

Flyvbjerg, B. (2001) *Making Social Science Matter: Why Social Inquiry Fails and How it Can Count Again*. New York: Cambridge University Press.

Fogg, B.J. and Eckles, D. (eds) (2007) *Mobile Persuasion: 20 Perspectives of the Future of Behavior Change*. Stanford: Captology Media.

Fontana, E.R. and Sørensen, C. (2005) 'From Idea to Blah! Understanding Mobile Services Development as Interactive Innovation', *Journal of Information Systems and Technology Management*, 2(2): 101–20.

Ford, J.D. and Backoff, R.W. (1988) 'Organizational Change in and out of Dualities and Paradox', in R.E. Quinn and K.S. Cameron (eds), *Paradox and Transformation: Toward a Theory of Change in Organization and Management*. Cambridge, MA: Ballinger, pp. 81–121.

Fortunati, L. (2002) 'The Mobile Phone: Towards New Categories and Social Relations', *Information, Communication & Society*, 5(4): 513–28.

——. (2005) 'Is Body-to-Body Communication Still the Prototype?', *Information Society*, 21: 53–61.

Fortunati, L., Katz, J.E. and Riccini, R. (2003) *Mediating the Human Body: Technology, Communication and Fashion*. Mahwah, NJ: Lawrence Erlbaum.

Fulton Suri, J. (2005) *Thoughtless Acts?* San Francisco: Chronicle Books.

The Future Foundation (2010) 'The Decisive Decade: How the Acceleration of Ideas will Transform the Workplace by 2020', survey of the future of work commisioned by Google.

Gambetta, D. and Hamill, H. (2005) *Streetwise: How Taxi Drivers Establish Customers' Trustworthiness*. London: Russell Sage Foundation.

Gant, D. and Kiesler, S. (2001) 'Blurring the Boundaries: Cell Phones, Mobility and the Line between Work and Personal Life', in B. Brown, N. Green and R. Harper (eds), *Wireless World*. Godalming: Springer-Verlag UK, pp. 121–31.

Garicano, L. and Heaton, P. (2010) 'Information Technology, Organization and Productivity in the Public Sector: Evidence from Police Departments', *Journal of Labor Economics*, 28(1): 167–201.

Gaver, W. (1991) 'Technology Affordances', in G.M. Olson and J.S. Olson (eds), *CHI '91: Proceedings of the SIGCHI Conference on Human Factors in Computing Systems: Reaching Through Technology*. New Orleans: ACM Press, pp. 79–84.

——. (1992) 'The Affordances of Media Spaces for Collaboration', in *Proceedings of the 1992 ACM Conference on Computer-Supported Cooperative Work*. New York: ACM Press, pp. 17–24.

Gawer, A. (ed.) (2009) *Platforms, Markets and Innovation*. Cheltenham: Edward Elgar.

Gergen, K.J. (2002) 'The Challenge of Absent Presence', in J.E. Katz and M. Aakhus (eds), *Perpetual Contact*. Cambridge University Press, pp. 227–41.

Gerson, E.M. and Star, S.L. (1986) 'Analyzing Due Process in the Workplace', *TOIS*, 4(3): 257–70.

Gersuny, C. and Rosengren, W.R. (1973) *The Service Society*. Cambridge, MA: Schenkman Publishing Co.

Gherardi, S. (2006) *Organizational Knowledge: The Texture of Workplace Learning*. Oxford: Blackwell.

Gibson, C. and Birkinshaw, J. (2004) 'The Antecedents, Consequences and Mediating Role of Organizational Ambidexterity', *Academy of Management Journal*, 47(2): 209–26.

Gibson, J.J. (1977) 'The Theory of Affordances', in R. Shaw and J. Bransford (eds), *The Theory of Affordances. In Perceiving, Acting and Knowing*. Chichester: John Wiley & Sons, pp. 67–82.

——. (1979) *The Ecological Approach to Visual Perception*. Chicago: Houghton Mifflin Harcourt.

Gibson, W. and Sterling, B. (1991) *The Difference Engine*. New York: Bantam Books.

Gioia, D. and Chittipeddi, K. (1991) 'Sensemaking and Sensegiving in Strategic Change Initiation', *Strategic Management Journal*, 12(6): 433–48.

Gladwell, M. (2008) *Outliers: The Story of Success*. London: Allen Lane.

Goffman, E. (1959) *The Presentation of Self in Everyday Life*. New York: Bantam Books.

——. (1971) *Relations in Public: Microstudies of the Public Order*. London: Allen Lane.

Golden, A.G. and Geisler, C. (2007) 'Work–Life Boundary Management and the Personal Digital Assistant', *Human Relations*, 60: 519–51.

Goldin, D.Q., Smolka, S.A. and Wegner, P. (eds) (2006) *Interactive Computation: The New Paradigm*. Berlin: Springer-Verlag.

González, V.M. and Mark, G. (2004) '"Constant, Constant, Multi-tasking Craziness": Managing Multiple Working Spheres', in E. Dykstra-Erickson and M. Tscheligi (eds), *Proceedings of ACM CHI 2004 Conference on Human Factors in Computing Systems*. New York: ACM Press, pp. 113–20.

Gotsi, M., Andriopoulos, C., Lewis, M.W. and Ingram, A.E. (2010) 'Managing Creatives: Paradoxical Approaches to Identity Regulation', *Human Relations*: 1–25.

Grandori, A. (1997) 'An Organizational Assessment of Interfirm Coordination Modes', *Organization Studies*, 18(6): 897–925.

Green, N. (2002) 'On the Move: Technology, Mobility and the Mediation of Social Time and Space', *Information Society*, 18: 281– 92.

Greenfield, A. (2006) *Everyware: The Dawning Age of Ubiquitous Computing*. Berkeley, CA: Peachpit Press.

Greeno, J.G. (1994) Gibson's Affordances. *Psychological Review*, 101(2): 336–42.

Gregory, D. and Urry, J. (eds) (1985) *Social Relations and Spatial Structures*. Basingtoke: Palgrave Macmillan.

Grier, D.A. (2005) *When Computers Were Human*. Princeton University Press.

Grönroos, C. (2000) *Service Management and Marketing – A Customer Relationship Management Approach*. Chichester: John Wiley & Sons.

Grudin, J. (2002) 'The Future of Business Services in the Age of Ubiquitous Computing', *Communications of the ACM*, 45(12): 74–8.

GSMA Mobile Infolink (2010) 'World Total Connections', http://www.gsmamobileinfolink. com, date accessed 6 May 2011.

Gutek, B. (1995) *The Dynamics of Service*. San Francisco: Jossey-Bass Wiley.

Haddon, L., Mante, E., Sapio, B., Kommonen, K.-H., Fortunati, L. and Kant, A. (eds) (2006) *Everyday Innovators: Researching the Role of Users in Shaping ICTs*. London: Springer.

Hägerstrand, T. (1975) 'Space, Time and Human Conditions', in A. Karlqvist, L. Lundqvist and F. Snickars (eds), *Dynamic Allocation of Urban Space*. Farnborough: Saxon House.

Hall, E. (1959) *The Silent Language*. New York: Doubleday.

——. (1962) *The Hidden Dimension*. New York: Anchor Press.

Halmos, P. (1970) *The Personal Service Society*. London: Constable.

Hambrick, D.C., Finkelstein, S. and Mooney, A.C. (2005) 'Executive Job Demands: New Insights for Explaining Strategic Decisions and Leader Behaviors', *Academy of Management Review*, 30(3): 472–91.

Handy, C. (1995) *The Age of Paradox*. Cambridge, MA: Harvard Business School Press.

Harper, R., Palen, L. and Taylor, A., ed. (2005) *The Inside Text: Social, Cultural and Design Perspectives on SMS*. Dordrecht: Springer.

Hayes, K. and Kuchinskas, S. (2003) *Going Mobile: Building the Real-time Enterprise with Mobile Applications that Work*. Berkeley, CA: CPM Books.

Heath, C. and Luff, P. (1992) 'Media Space and Communicative Asymmetries: Preliminary Observations of Video-Mediated Interaction', *Human-Computer Interaction*, 7: 315–46.

——. (2000) *Technology in Action*. Cambridge University Press.

Heide, L. (2005) *Hulkort og EDB i Danmark 1911–1970*. Århus: Systime.

Herskind, S. (1996) 'Truck Drivers Know Where to Go and When to be There – or do They?', in B. Dahlbom, F. Ljungberg, U. Nuldén, K. Simon, C. Sørensen and J. Stage (eds), *The 19th Information Systems Research Seminar in Scandinavia*. Gothenburg: Gothenburg Studies in Informatics, pp. 231–46.

Herzhoff, J., Elaluf-Calderwood, S. and Sørensen, C. (2010) 'Convergence, Conflicts and Control Points: A Systems-Theoretical Analysis of Mobile VoIP in the UK', in George M. Giaglis and Nikos Mylonopoulos (eds), *Proceedings of Joint 9th International Conference on Mobile Business (ICMB 2010) and 9th Global Mobility Roundtable (GMR 2010). Athens, Greece*. Athens: IEEE Computer Society Press.

Hiltz, S.R. and Turoff, M. (1985) 'Structuring Computer-Mediated Communication Systems to Avoid Information Overload', *Communications of the ACM*, 28(7): 680–9.

Hinds, P.J. and Kiesler, S. (1995) 'Communication across Boundaries: Work, Structure and Use of Communication Technologies in a Large Organization', *Organization Science*, 6(4): 373–93.

——. (ed.) (2002) *Distributed Work*. Cambridge, MA: MIT Press.

Hislop, D. (ed.) (2008) *Mobility and Technology in the Workplace*. London: Routledge.

Hislop, D. and Axtell, C. (2007) 'The Neglect of Spatial Mobility in Contemporary Studies of Work: The Case of Telework', *New Technology, Work and Employment*, 22(1): 34–51.

Hjorth, L. (2009) *Mobile Media in the Asia-Pacific: The Art of Being Mobile*. Oxford: Routledge.

Hochschild, A.R. (1997) *The Time Bind: When Work Becomes Home and Home Becomes Work*. New York: Owl Books.

Holman, D., Wall, T.D., Clegg, C.W., Sparrow, P. and Howard, A. (eds) (2003) *The Essentials of the New Workplace: A Guide to the Human Impact of Modern Working Practices*. Chichester: Wiley.

Holmberg, L. and Mathiassen, L. (2001) 'Survival Patterns in Fast-Moving Software Organizations', *IEEE Software*, 18(6): 51–5.

Hong, S.-J. and Tam, K.Y. (2006) 'Understanding the Adoption of Multipurpose Information Appliances: The Case of Mobile Data Services', *Information Systems Research*, 17(2): 162–79.

Horst, H. and Miller, D. (2006) *The Cell Phone: An Anthropology of Communication*. New York: Berg.

Hughes, J., O'Brien, J., Randall, D., Rouncefield, M. and Tolmie, P. (2001) 'Some "Real" Problems of "Virtual" Organisation', *New Technology, Work and Employment*, 16(1): 49–64.

Hughes, J.A., Rouncefield, M. and Tolmie, P. (2002) 'The Day-to-Day Work of Standardization: A Sceptical Note on the Reliance on IT in a Retail Bank', in S. Woelgar (ed.), *Virtual Society?: Technology, Cyberbole, Reality*. Oxford University Press, pp. 247–63.

Hutchby, I. (2001) 'Technologies, Texts and Affordances', *Sociology*, 35(2): 441–56.

Hutchins, E. (1995) *Cognition in the Wild*. Cambridge, MA: MIT Press.

Iastrebova, K. (2006) 'Managers' Information Overload: The Impact of Coping Strategies on Decision-Making Performance', PhD dissertation, Erasmus University.

Ito, M., Okabe, D. and Matsuda, M. (ed.) (2005) *Persona, Portable, Pedestrian: Mobile Phones in Japanese Life*. Cambridge, MA: MIT Press.

ITU (2009) 'Measuring the Information Society: The ICT Development Index. International Telecommunications Union', http://www.itu.int/ITU-D/ict/publications/idi/2009/material/IDI2009_w5.pdf, date accessed 6 May 2011.

Jackson, M. (2008) *Distracted: The Erosion of Attention in the Coming Dark Age*. New York: Prometheus Books.

Jackson, P.J. and van der Wielen, J.M. (eds) (1998) *Teleworking: International Perspectives – From Telecommuting to the Virtual Organisation*. London: Routledge.

Jacoby, J. (1984) 'Perspectives on Information Overload', *Journal of Consumer Research*, 10(4): 432–5.

Jarvenpaa, S.L. and Lang, K.R. (2005) 'Managing the Paradoxes of Mobile Technology', *Information Systems Management*, 22(4): 7–23.

Jarvenpaa, S.L. and Leidner, D.E. (1999) 'Communication and Trust in Global Virtual Teams', *Organization Science*, 10(6): 791–815.

Jessup, L.M. and Robey, D. (2002) 'The Relevance of Social Issues in Ubiquitous Computing Environments', *Communications of the ACM*, 45(12): 88–91.

Johansen, R. (1988) *Groupware: Computer Support for Business Teams*. New York: Free Press.

Kakihara, M. (2003) 'Emerging Work Practices of ICT-Enabled Mobile Professionals', PhD dissertation, London School of Economics and Political Science.

Kakihara, M. and Sørensen, C. (2001) 'Expanding the "Mobility" Concept', *ACM SIGGROUP Bulletin*, 22(3): 33–7.

——. (2002) 'Mobility: An Extended Perspective', in R. Sprague Jr. (ed.), *Thirty-Fifth Hawaii International Conference on System Sciences (HICSS-35)*. Big Island Hawaii: IEEE.

——. (2004) 'Practicing Mobile Professional Work: Tales of Locational, Operational and Interactional Mobility', *INFO: The Journal of Policy, Regulation and Strategy for Telecommunication, Information and Media*, 6(3): 180–7.

Kakihara, M., Sørensen, C. and Wiberg, M. (2005) 'Negotiating the Fluidity of Mobile Work', in M. Wiburg (ed.), *The Interaction Society: Practice, Theories, & Supportive Technologies*. Hershey, PA: Idea Group Inc., Chapter 7.

Kallinikos, J. (1996) *Technology and Society*. Munich: Accedo.

——. (2006) *The Consequences of Information: Institutional Implications of Technological Change*. Cheltenham: Edward Elgar.

Kane, T. (2003) 'The Last Ten Years: Developments in the Foreign Exchange Market – A Broker's Perspective', in F. Taylor (ed.), *Mastering Foreign Exchange and Currency Options: A Practitioner's Guide to the New Marketplace*. London: Prentice Hall.

Karsten, H. (2003) 'Constructing Interdependencies with Collaborative Information Technology', *Computer Supported Cooperative Work*, 12: 437–64.

Kellaway, L. (2005) *Martin Lukes: Who Moved My Blackberry?* London: Penguin.

Kietzmann, J. (2007) 'In Touch out in the Field: Coalescence and Interactive Innovation of Technology for Mobile Work', PhD dissertation, London School of Economics and Political Science.

——. (2008) 'Internative Innovation of Technology for Mobile Work', *European Journal of Information Systems*, 17(3): 305–20.

Kilduff, M., Funk, J. and Mehra, A. (1997) 'Engineering Identity in a Japanese Factory', *Organization Science*, 8(6): 579–92.

Kjeldskov, J. and Graham, C. (2003) 'A Review of Mobile HCI Research Methods', in *Proceedings of Mobile HCI 2003*. Udine, Italy: ACM Press.

Kjeldskov, J. and Stage, J. (2004) 'New Techniques for Usability Evaluation of Mobile Systems', *International Journal of Human-Computer Studies*, 60: 599–620.

Kleinman, S. (ed.) (2007) *Displacing Place: Mobile Communication in the Twenty-First Century*. New York: Peter Lang.

Kleinrock, L. (1996) 'Nomadicity: Anytime, Anywhere in a Disconnected World', *Mobile Networks and Applications*, 1: 351–7.

Klockars, C.B. (1985) *The Idea of Police*. Beverly Hills, CA: Sage.

Kluth, A. (2008) 'Nomads at Last – A Special Report on Mobile Telecoms', *The Economist*, 12 April, http://www.economist.com/specialreports/displayStory.cfm?story_id=10950394, date accessed 6 May 2011.

Knorr-Cetina, K. and Bruegger, U. (2002) 'Global Microstructures: The Virtual Societies of Financial Markets', *American Journal of Sociology*, 107(4): 905–50.

Kodres, L.E. (1996) 'Foreign Exchange Markets: Structure and Systemic Risks', World Bank.

Kopomaa, T. (2000) *The City in Your Pocket: Birth of the Mobile Information Society*, T. Snellman (trans). Helsinki: Gaudeamus.

Kotlarsky, J., Oshri, I. and van Fenema, P. (2008) *Knowledge Processes in Globally Distributed Contexts*. Basingstoke: Palgrave Macmillan.

Kourouthanassis, P.E. and Giaglis, G.M. (eds) (2008) *Pervasive Information Systems*. Armonk, NY: M.E. Sharpe.

Kourouthanassis, P.E., Giaglis, G.M. and Karaiskos, D.C. (2010) 'Delineating "Pervasiveness" in Pervasive Information Systems: A Taxonomical Framework and Design Implications', *Journal of Information Technology*, 25(3): 273–87.

Kraut, R.E. and Streeter, L.A. (1995) 'Coordination in Software Development', *Communications of the ACM*, 38(3): 69–81.

Kristoffersen, S. and Ljungberg, F. (1999) 'Making Place to Make IT Work: Empirical Explorations of HCI for Mobile CSCW', in *Proceedings of the International Conference on Supporting Group Work (GROUP '99), Phoenix, Arizona, United States*. New York: ACM Press, pp. 276–85.

——. (2000) 'Mobility: From Stationary to Mobile Work', in K. Braa, C. Sørensen and B. Dahlbom. *Planet Internet*. Lund, Sweden: Studentliteratur, pp. 41–64.

Krogstie, J., Kautz, K. and Allen, D. (eds) (2005) *Mobile Information Systems II*. Berlin: Springer-Verlag.

Kunda, G. (1992) *Engineering Culture: Control and Commitment in a High-Tech Corporation*. Philadelphia: Temple University Press.

Kurke, L. and Aldrich, H. (1983) 'Mintzberg was Right!: A Replication and Extension of the Nature of Managerial Work', *Management Science*, 29(8): 975–84.

Laing, G. (2004) *Digital Retro: The Evolution and Design of the Personal Computer*. Lewes: Ilex.

Laing, A., Duffy, F. and Jaunzens, D. (1998) *New Environments for Working: The Re-design of Offices and Environmental Systems for New Ways of Working*. Watford: Construction Research Communications Ltd.

Lamond, D., Daniels, K. and Standen, P. (2003) 'Managing Virtual Workers and Virtual Organisations', in D. Holman, T. Wall, C. Clegg, P. Sparrow and A. Howard (eds), *The Essentials of the New Workplace: A Guide to the Human Impact of Modern Working Practices*. Chichester: Wiley, pp. 173–95.

Landau, J. (2010) 'The Conceptualization of Practice in IS Research on Mobile ICT', MSc dissertation, London School of Economics.

Lattanzi, M., Korhonen, A. and Gopalakrishnan, V. (2006) *Work Goes Mobile: Nokia's Lessons from the Leading Edge*. Chichester: Wiley.

Laurier, E. (2004) 'Doing Office Work on the Motorway', *Theory, Culture & Society*, 21(4/5): 261–77.

Lawton, G. (2004) 'Machine-to-Machine Technology Gears up for Growth', *Computer*, 37(9): 12–15.

Leavitt, H.J. (1964) 'Applied Organization Change in Industry: Structural, Technical and Human Approaches', in W. Cooper, H. Leavitt and M. Shelly II (eds), *New Perspectives in Organization Research*. New York: John Wiley & Sons, pp. 55–71.

Lee, H. (1999) 'Time and Information Technology: Monochronicity, Polychronicity and Temporal Symmetry', *European Journal of Information Systems*, 8(1): 16–26.

Lee, H. and Liebenau, J. (2002) 'A New Time Discipline: Managing Virtual Work Environments', in R. Whipp, B. Adam and I. Sabelis (eds), *Making Time: Time and Management in Modern Organizations*. Oxford University Press, pp. 115–25.

Lefebvre, H. (1991) *The Production of Space*. Oxford: Blackwell.

——. (2004) *Rhythmanalysis: Space, Time and Everyday Life*. New York: Continuum.

Leifer, R. (1988) 'Matching Computer-Based Information Systems with Organizational Structures', *MIS Quarterly*, 12: 63–73.

Leinonen, P., Järvelä, S. and Häkkinen, P. (2005) 'Conceptualizing the Awareness of Collaboration: A Qualitative Study of a Global Virtual Team', *Computer Supported Cooperative Work*, 14(4): 301–22.

Leonardi, P.M. (2011) 'When Flexible Routines Meet Flexible Technologies: Affordance, Constraint and the Imbrication of Human and Material Agencies', *MIS Quarterly*, 35(1): 147–67.

Leonardi, P.M. and Barley, S.R. (2010) 'What's Under Construction Here? Social Action, Materiality and Power in Constructivist Studies of Technology and Organizing', *Academy of Management Annals*, 4(1): 1–51.

Lewis, M. (2000) 'Exploring Paradox: Toward a More Comprehensive Guide', *Academy of Management Review*, 25(4): 760–76.

Lewis, M., Welsh, M., Dehler, G. and Green, S. (2002) 'Product Development Tensions: Exploring Contrasting Styles of Project Management', *Academy of Management Journal*, 45(3): 546–64.

Liao, Z. (2003) 'Real-Time Taxi Dispatching using Global Positioning Systems', *Communications of the ACM*, 46(5): 81–3.

Licoppe, C. (2004) '"Connected" Presence: The Emergence of a New Repertoire for Managing Social Relationships in a Changing Communication Technoscape', *Environment and Planning D: Society and Space*, 22: 135–56.

Licoppe, C. and Smoreda, Z. (2006) 'Rhythms and Ties: Towards a Pragmatics of Technologically-Mediated Sociability', in R. Kraut, M. Brynin and S. Kiesler (eds), *Computers, Phones and the Internet: Domesticating Information Technology*. New York: Oxford University Press.

Ling, R. (2004) *The Mobile Connection: The Cell Phone's Impact on Society*. Amsterdam: Morgan Kaufmann.

——. (2008) *New Tech, New Ties: How Mobile Communication is Reshaping Social Cohesion*. Cambridge, MA: MIT Press.

Ljungberg, F. (1997) 'Networking', PhD dissertation, Gothenburg University.

——. (1999) 'Exploring CSCW Mechanisms to Realize Constant Accessibility without Inappropriate Interaction', *Scandinavian Journal of Information Systems*, 11(1): 25–50.

Ljungberg, F. and Sørensen, C. (2000) 'Overload: From Transaction to Interaction', in K. Braa, C. Sørensen and B. Dahlbom (eds), *Planet Internet*. Lund, Sweden: Studentlitteratur, pp. 113–36.

Love, S. (2005) *Understanding Mobile Human-Computer Interaction: A Psychological Perspective*. Oxford; Butterworth-Heinemann.

——. (2009) *Handbook of Mobile Technology Research Methods*. Hauppauge, NY: Nova Science Publishers Inc.

Lovelock, C. (1983) 'Classifying Services to Gain Strategic Marketing Insights', *Journal or Marketing*, 47(3): 9–20.

Luff, P. and Heath, C. (1998) 'Mobility in Collaboration', in *Proceedings of ACM 1998 Conference on Computer Supported Cooperative Work*. New York: ACM Press.

Lüscher, L. and Lewis, M. (2008) 'Organizational Change and Managerial Sensemaking: Working Through Paradox', *Academy of Management Journal*, 51(2): 221–40.

Lyon, D. (2009) *Identifying Citizens*. Cambridge: Polity Press.

Lyytinen, K. (2003) 'The Next Wave of IS Research: Design and Investigation of Ubiquitous Computing', panel description in C. Ciborra, R. Mercurio, M. De Marco, M. Martinez and A. Carignani (eds), *Proceedings of the 11th European Conference on Information Systems, ECIS 2003, Naples, Italy June 16–21*. University of Naples.

Lyytinen, K. and Yoo, Y. (2002a) 'Issues and Challenges in Ubiquitous Computing', *Communications of the ACM*, 45(12): 62–5.

——. (2002b) 'The Next Wave of Nomadic Computing: A Research Agenda for Information Systems Research', *Information Systems Research*, 13(4): 377–88.

Mackay, W.E. (1988) 'Diversity in the Use of Electronic Mail: A Preliminary Inquiry', *TOIS: ACM Transactions on Office Information Systems*, 6(4): 380–97.

——. (2000) 'Responding to Cognitive Overload: Co-adaptation Between Users and Technology', *Intellectica*, 30(1): 177–93.

Maes, P. (1991) *Designing Autonomous Agents: Theory and Practice from Biology to Engineering and Back*. Cambridge. MA: MIT Press.

——. (1994) 'Agents that Reduce Work and Information Overload', *Communications of the ACM*, 37(7): 31–40.

Makimoto, T. and Manners, D. (1997) *Digital Nomad*. Chichester: Wiley.

Malhotra, N. (1984) 'Reflections on the Information Overload Paradigm in Consumer Decision Making', *Journal of Consumer Research*, 10(4): 436–40.

Malone, T.W. (2004) *The Future of Work: How the New Order of Business Will Shape Your Organization, Your Management Style and Your Life*. Harvard Business School Press.

Malone, T.W. and Crowston, K. (2001) 'The Interdisciplinary Study of Coordination', in G. Olson, T. Malone and J. Smith (eds), *Coordination Theory and Collaboration Technology*. Hillsdale, NJ: Lawrence Erlbaum, pp. 7–50.

Malone, T.W., Grant, K.R., Lai, K.-Y., Rao, R. and Rosenblitt, D. (1987) 'Semistructured Messages are Surprisingly Useful for Computer-Supported Coordination', *TOIS*, 5(2): 115–31.

Malone, T.W. and Laubacher, R.J. (1998) 'The Dawn of the E-Lance Economy', *Harvard Business Review* 145–53.

Mann, S. and Niedzviecki, H. (2002) *Cyborg: Digital Destiny and Human Possibility in the Age of the Wearable Computer*. New York: Doubleday.

Manning, P.K. (2003) *Policing Contingencies*. University of Chicago Press.

——. (2008) *The Technology of Policing: Crime Mapping, Information Technology and the Rationality of Crime Control*. New York University Press.

Mansell, R., Avgerou, C., Quah, D. and Silverstone, R. (eds) (2007) *The Oxford Handbook of Information and Communication Technologies*. Oxford University Press.

Manwaring-White, S. (1983) *The Policing Revolution: Police Technology, Democracy and Liberty in Britain*. Brighton: Harvester.

March, S., Raghu, T.S. and Vinze, A. (2008) 'Editorial Introduction: Cultivating and Securing the Information Supply Chain', *Journal of the Association for Information Systems*, 9(3/4): 95–7.

Markus, L. (1994) 'Electronic Mail as the Medium of Managerial Choice', *Organization Science*, 5(4): 502–27.

Markus, M.L. and Robey, D. (2004) 'Why Stuff Happens: Explaining the Unintended Consequences of Using IT', in K. Andersen and M. Vendelø (eds), *The Past and Future of Information Systems*. Oxford: Butterworth-Heinemann.

Martinez, J.I. and Jarillo, J.C. (1989) 'The Evolution of Research on Coordination Mechanisms in Multinational Corporations', *Journal of International Business Studies*, 20(3): 489–514.

Massey, D. and Jess, P.M. (eds) (1995) *A Place in the World? Places, Cultures and Globalization*. Oxford University Press.

Mathiassen, L. (2010) 'Intended Affordances and Protocols versus Performances', personal communication during HICSS, Hawaii.

Mathiassen, L., Munk-Madsen, A., Nielsen, P.A. and Stage, J. (2000) *Object Oriented Analysis and Design*. Aalborg: Marko Publishing.

Mathiassen, L. and Sørensen, C. (2008) 'Towards a Theory of Organizational Information Services', *Journal of Information Technology*, 23(4): 313–29.

Mayer-Schonberger, V. (2009) *Delete: The Virtue of Forgetting in the Digital Age*. Princeton University Press.

Mazmanian, M., Yates, J. and Orlikowski, W. (2006) 'Ubiquitous Email: Individual Experiences and Organizational Consequences of Blackberry Use', in *Proceedings of the 65th Annual Meeting of the Academy of Management*, Atlanta.

Mazmanian, M.A., Orlikowski, W.J. and Yates, J. (2005) 'Crackberries: The Social Implications of Ubiquitous Wireless E-Mail Devices', in C. Sørensen, Y. Yoo, K. Lyytinen and J. DeGross (eds), *Designing Ubiquitous Information Environments: Sociotechnical Issues and Challenges*. New York: Springer, pp. 337–43.

McCollough, M. (2004) *Digital Ground: Architecture, Pervasive Computing and Environmental Knowing*. Cambridge, MA: MIT Press.

McFarlane, D.C. (2002) 'Comparison of Four Primary Methods for Coordinating the Interruption of People in Human–Computer Interaction', *Human-Computer Interaction*, 17: 63–139.

McFarlane, D.C. and Latorella, K.A. (2002) 'The Scope and Importance of Human Interruption in Human–Computer Interaction Design', *Human-Computer Interaction*, 17: 1–61.

McGrath, J.E. (1991) 'Time, Interaction and Performace (Tip): A Theory of Groups', *Small Group Research*, 22(2): 147–74.

McGrenere, J. and Ho, W. (2000) 'Affordances: Clarifying and Evolving a Concept', *Graphics Interface*: 179–86.

McGuire, R. (2007) *The Power of Mobility: How Your Business Can Compete and Win in the Next Revolution*. Hoboken, NJ: John Wiley & Sons.

Mick, D.G. and Fournier, S. (1998) 'Paradoxes of Technology: Consumer Cognizance, Emotions and Coping Strategies', *Journal of Consumer Research*, 25: 123–43.

Mills, C.W. (1951) *White Collar. The American Middle Classes*. New York: Oxford University Press.

Mintzberg, H. (1971) 'Managerial Work: Analysis from Observation', *Management Science*, 18(2): B97–B110.

——. (1983) *Structure in Fives: Designing Effective Organizations*. Englewood Cliffs, NJ: Prentice Hall.

Mitchell, W.J. (2003) *Me++: The Cyborg Self and the Networked City*. Cambridge, MA: MIT Press.

Mol, A. and Law, J. (1994) 'Regions, Networks and Fluids: Anaemia and Social Topology', *Social Studies of Science*, 24: 641–71.

Mom, T.J.M., Van den Bosch, F.A.J. and Volberda, H.W. (2009) 'Understanding Variation in Managers' Ambidexterity: Investigating Direct and Interaction Effects of Formal Structural and Personal Coordination Mechanisms', *Organization Science*, 20(4): 812–28.

Morley, D. (2000) *Home Territories: Media, Mobility and Identity*. London: Routledge.

Morus, I.R. (ed.) (2002) *Bodies/Machines*. New York: Berg.

Myerson, J. and Ross, P. (2006) *Space to Work: New Office Design*. London: Lawrence King Publishing.

Naisbitt, J. (1994) *Global Paradox*. New York: Avon Books.

Nandhakumar, J. (2002) 'Managing Time in a Software Factory: Temporal and Spatial Organization of IS Development Activities', *Information Society*, 18(4): 251–62.

Nardi, B. and Whittaker, S. (2000) 'Interaction and Outeraction', in W. Kellogg and S. Whittaker (eds), *Proceedings of Computer Supported Cooperative Work*. Philadelphia: ACM Press, pp. 79–88.

Nardi, B. and O'Day, V. (1999) *Information Ecologies: Using Technology with Heart*. Cambridge, MA: MIT Press.

Nardi, B., Whittaker, S. and Schwarz, H. (2002) 'NetWORKers and their Activity in Intensional Networks', *Computer Supported Cooperative Work*, 11: 205–42.

Newell, S., Robertson, M., Scarbrough, H., and Swan, J. (2009) *Managing Knowledge Work amd Innovation*. Basingstoke: Palgrave Macmillan.

Newman, W. (1994) 'A Preliminary Analysis of the Products of HCI Research Using Pro Forma Abstracts', in *Proceedings of the CHI'9 4 Conference on Computer–Human Interaction*. New York: ACM Press, pp. 278–84.

Ngwenyama, O.K. and Lee, A.S. (1997) 'Communication Richness in Electronic Mail: Critical Theory and the Contextuality of Meaning', *MIS Quarterly*, 21(2): 147–67.

Nilles, J.M. (1998) *Managing Telework: Strategies for Managing the Virtual Workforce*. New York: John Wiley & Sons.

Nonaka, I. and Konno, N. (1998) 'The Concept of "Ba": Building a Foundation for Knowledge Creation', *California Management Review*, 40(3): 40–54.

Norman, D. (1988) *The Psychology of Everyday Things*. New York: Basic Books.

——. (1999) 'Affordance, Conventions and Design', *Interactions*, 6(3): 38–42.

O'Conaill, B. and Frohlich, D. (1995) 'Timespace in the Workplace: Dealing with Interruptions', in I. Katz, R. Mack and L. Marks (eds), *Proceedings of CHI'95: Human Factors in Computing Systems*. Denver, CO: ACM Press, pp. 262–3.

O'Reilly, C.A. and Tushman, M.L. (2004) 'The Ambidextrous Organization', *Harvard Business Review*, pp. 74–81.

Oard, D.W. (1997) 'The State of the Art in Text Filtering', *User Modeling and User-Adapted Interaction: An International Journal*, 7(3): 141–78.

Oliveira, A. (2006) 'Connected Adventure: An Examination of the Use of ICTs in Expeditions to Support a Collective Experience', MSc dissertation, Department of Management, Information Systems and Innovation Group, London School of Economics.

Olson, G.M. and Olson, J.S. (2000) 'Distance Matters', *Human-Computer Interaction*, 15: 139–78.

Orlikowski, W. (2007) 'Sociomaterial Practices: Exploring Technology at Work', *Organization Science*, 28(9): 1435–48.

Orlikowski, W. and Iacono, C.S. (2001) 'Research Commentary: Desperately Seeking the "IT" in IT Research – A Call to Theorizing the IT Artifact', *Information Systems Research*, 12(2): 121–34.

Orlikowski, W. and Scott, S.V. (2008) 'Sociomateriality: Challenging the Separation of Technology, Work and Organization', *Academy of Management Annals*, 2(1): 433–74.

Orr, J.E. (1996) *Talking About Machines: An Ethnography of a Modern Job*. New York: Cornell University Press.

Palme, J. (1984) 'You Have 134 Unread Mail! Do You Want to Read Them Now?', in H.T. Smith (ed.), *Computer-Based Message Semites: IFIP WG 6.5 Working Conference on Computer-based Document Services*. New York: Elsevier North-Holland, pp. 175–84.

Pearlson, K. and Saunders, C. (2001) 'There's No Place Like Home: Managing Telecommuting Paradoxes', *Academy of Management Executive (1993–2005)*, 15(2): 117–28.

Peppard, J. (2003) 'Managing IT as a Portfolio of Services', *European Management Journal*, 21(4): 467–83.

Perkin, H.J. (2002) *The Rise of Professional Society: England since 1880*. London: Routledge.

Perlow, L. (1998) 'Boundary Control: The Social Ordering of Work and Family Time in a High-Tech Corporation', *Administrative Science Quarterly*, 43(2): 328–57.

Perlow, L., Hoffer Gittell, J. and Katz, N. (2004) 'Contextualizing Patterns of Work Group Interaction: Toward a Nested Theory of Structuration', *Organization Science*, 15(5): 520–36.

Perrow, C. (1984) *Normal Accidents. Living with High-Risk Technologies*. New York: Basic Books.

Perry, M., O'Hara, K., Sellen, A., Brown, B. and Harper, R. (2001) 'Dealing with Mobility: Understanding Access Anytime, Anywhere', *ACM Transactions on Computer-Human Interaction*, 8(4): 323–47.

Pertierra, R. (ed.) (2007) *The Social Construction and Usage of Communication Technologies: Asian and European Experiences*. Quezon City: University of the Philippines Press.

Peters, P.F. (2005) *Time Innovation and Mobilities*. London: Routledge.

Peters, T. and Waterman Jr., R.H. (2004, first edn 1982) *In Search of Excellence: Lessons from America's Best-Run Companies*. London: Profile Books.

Pica, D. (2006) 'The Rhythms of Interaction with Mobile Technologies: Tales from the Police', PhD dissertation, London School of Economics.

Pica, D. and Kakihara, M. (2003) 'The Duality of Mobility: Understanding Fluid Organizations and Stable Interaction', in C. Ciborra, R. Mercurio, M. De Marco, M. Martinez and A. Carignani (eds), *Proceedings of the 11th European Conference on Information Systems, ECIS 2003, Naples, Italy June 16–21*. University of Naples.

Picard, R.W. (1997) *Affective Computing*. Cambridge, MA: MIT Press.

Pogue, D. (2009) 'Pogue's Post: Some E-Books Are More Equal than Others', *New York Times*, 17 July, http://pogue.blogs.nytimes.com/2009/07/17/some-e-books-are-more-equal-than-others/, date accessed 6 May 2011.

Poole, M.S. and Van de Ven, A.H. (1989) 'Using Paradox to Build Management and Organization Theories', *Academy of Management Review*, 14(4): 562–78.

Pooley, C.G., Turnbull, J. and Adams, M. (2005) *A Mobile Century? Changes in Everyday Mobility in Britain in the Twentieth Century*. Aldershot: Ashgate.

Postman, N. (1992) *Technopoly: The Surrender of Culture to Technology*. New York: Vintage Books.

Prasad, P. and Prasad, A. (2000) 'Stretching the Iron Cage: The Constitution and Implications of Routine Workplace Resistance', *Organization Science*, 11(4): 387–403.

Putnam, R.D. (2000) *Bowling Alone*. New York: Touchstone.

Qui, A. (2006) 'Knowledge Management Issues in a Professional Accounting Firm', MSc dissertation, London School of Economics.

Quinn, R. and Rohrbaugh, J. (1983) 'A Spatial Model of Effectiveness Criteria: Towards a Competing Values Approach to Organizational Analysis', *Management Science*, 29(3): 363–77.

Raisch, S., Birkinshaw, J., Probst, G. and Tushman, M.L. (2009) 'Organizational Ambidexterity: Balancing Exploitation and Exploration for Sustained Performance', *Organization Science*, 20(4): 685–95.

Ramaprasad, A. and Rai, A. (1996) 'Envisioning Management of Information', *Omega, International Journal of Management Science*, 24(2): 179–93.

Randall, D., Harper, R. and Rouncefield, M. (2007) *Fieldwork for Design: Theory and Practice*. London: Springer.

Rao, M. and Mendoza, L. (eds) (2004) *Asia Unplugged: The Wireless and Mobile Media Boom in Asia-Pacific*. London: Sage.

Rasmussen, J., Pejtersen, A.M. and Goodstein, L.P. (1994) *Cognitive Systems Engineering*. New York: John Wiley & Sons.

Reid, N. (2010) *Wireless Mobility*. New York: McGraw-Hill.

Rettie, R. (2009) 'Mobile Phone Communication: Extending Goffman to Mediated Interaction', *Sociology*, 43(3): 421–38.

Richards, M. and Alderman, J. (2007) *Core Memory: A Visual Survey of Vintage Computers*. San Francisco: Chronicle Books.

Robertson, M. and Swan, J. (2003) '"Control – What Control?" Culture and Ambiguity Within a Knowledge Intensive Firm', *Journal of Management Studies*, 40(4): 831–59.

Robey, D. and Boudreau, M.-C. (1999) 'Accounting for the Contradictory Organizational Consequences of Information Technology: Theoretical Directions and Methodological Implications', *Information Systems Research*, 10(2): 167–85.

Rosenwald, M.S. (2009) 'Digital Nomads Choose their Tribes: Teleworkers Find Camaraderie in a New Kind of Colleague', *Washington Post*, 25 July, http://www.washingtonpost.com/wp-dyn/content/article/2009/07/25/AR2009072500878_pf.html, date accessed 6 May 2011.

Rouncefield, M., Hughes, J.A., Rodden, T. and Viller, S. (1994) 'Working with "Constant Interruption": CSCW and the Small Office', in T. Malone (ed.), *CSCW '94. Proceedings of the Conference on Computer-Supported Cooperative Work, Chapel Hill, North Carolina, October 24–26*. New York: ACM Press, pp. 275–86.

Rousseau, D.M., Sitkin, S.B., Burt, R.S. and Camerer, C. (1998) 'Not So Different After All: A Crossdiscipline View of Trust', *Academy of Management Review*, 23(3): 393–404.

Ryan, S., Jaffe, J., Drake, S.D. and Boggs, R. (2009) 'Worldwide Mobile Worker Population 2009–2013 Forecast', IDC, http://www.idc.com/getdoc.jsp?containerId=221309, date accessed 6 May 2011.

Saccol, A.Z. and Reinhard, N. (2006) 'The Hospitality Metaphor as a Theoretical Lens for Understanding the ICT Adoption Process', *Journal of Information Technology*, 21: 154–64.

Sacks, H., Schegloff, E.A. and Jefferson, G. (1974) 'A Simplest Systematics for the Organization of Turn-Taking for Conversation', *Language*, 50(4): 696–735.

Sarnobat, A. (2006) 'Activity Theory and the Conceptualization and Analysis of Design', MSc dissertation, London School of Economics.

Sawhney, N. and Schmandt, C. (2000) 'Nomadic Radio: Speech and Audio Interaction for Contextual Messaging in Nomadic Environments', *ACM Transactions of Computer-Human Interaction*, 7(3): 353–83.

Sawyer, S. and Tapia, A. (2006) 'The Sociotechnical Nature of Mobile Computing Work: Evidence from a Study of Policing in the United States', in B. Stahl (ed.), *Issues and Trends in Technology and Human Interaction*. Hershey, PA: Idea Group Publishing, pp. 152–71.

Scarbrough, H. (1995) 'Blackboxes, Hostages and Prisoners', *Organization Studies*, 16(6): 991–1019.

Schmidt, K. (1993) 'Modes and Mechanisms of Interaction in Cooperative Work', in C. Simone and K. Schmidt (eds), *Computational Mechanisms of Interaction for CSCW*. University of Lancaster Press, pp. 21–104.

——. (1994) 'The Organization of Cooperative Work — Beyond the 'Leviathan' Conception of the Organization of Cooperative Work', in T. Malone (ed.), *CSCW '94. Proceedings of the Conference on Computer-Supported Cooperative Work, Chapel Hill, North Carolina, October 24–26, 1994*. New York: ACM Press, pp. 101–12.

Schmidt, K. (1999) 'Of Maps and Scripts: The Status of Formal Constructs in Cooperative Work', *Information and Software Technology*, 41(6): 319–29.

Schmidt, K. (2002) 'The Problem with "Awareness": Introductory Remarks on "Awareness in CSCW"', *Computer Supported Cooperative Work*, 11(3–4): 285–98.

Schmidt, K. and Bannon, L. (1992) 'Taking CSCW Seriously: Supporting Articulation Work', *CSCW Journal*, 1(1–2): 7–40.

Schmidt, K. and Simone, C. (1996) 'Coordination Mechanisms: Towards a Conceptual Foundation of CSCW Systems Design', *Computer Supported Cooperative Work: The Journal of Collaborative Computing*, 5(2–3): 155–200.

Schneiderman, B. and Maes, P. (1997) 'Direct Manipulation vs Software Agents: Excerpts from Debates at IUI 97 and CHI 97', *Interactions*, 4(6): 42–61.

Schultze, U. and Vandenbosch, B. (1998) 'Information Overload in a Groupware Environment: Now You See it, Now You Don't', *Journal of Organizational Computing and Electronic Commerce*, 8(2): 127–48.

Schwager, J.D. (1992) *The New Market Wizards. Conversations with America's Top Traders*. New York: HarperCollins.

Scornavacca, E. and Barnes, S.J. (2008) 'The Strategic Value of Enterprise Mobility: Case Study Insights', *Information Knowledge Systems Management Journal*, 7(1 and 2): 227–41.

Sellen, A.J. and Harper, R. (2003) *The Myth of the Paperless Office*. Cambridge, MA: MIT Press.

Sennett, R. (2008) *The Craftsman*. London: Allen Lane.

Sheedy, T. (2010) 'Insights For CIOs: Make Mobility Standard Business Practice'. Forrester Research.

Sherry, J. and Salvador, T. (2001) 'Running and Grimacing: The Struggle for Balance in Mobile Work', in B. Brown, N. Green and R. Harper (eds), *Wireless World*. Godalming: Springer-Verlag UK, pp. 108–20.

Sidorova, A., Evangelopoulos, N., Valacich, J.S. and Ramakrishnan, T. (2008) 'Uncovering the Intellectual Core of the Information Systems Discipline', *MIS Quarterly*, 32(3): 467–82.

Singh, R., Mathiassen, L. and Mishra, A.N. (2009) 'A Theory of Rural Telehealth: A Paradoxical Analysis of Effective Innovation Paths', in Proceedings of the International Conference for Information Systems, *ICIS 2009. Paper 126*. Association of Information Systems, Atlanta, Georgia. http://aisel.aisnet.org/icis2009/126, date accessed 17 May 2011.

Skok, W. and Baird, S. (2005) 'Strategic Use of Emerging Technology in the Taxi Cab Industry', *Knowledge and Process Management*, 14: 295–306.

Skok, W. and Kobayashi, S. (2005) 'An International Taxicab Evaluation: Comparing Tokyo with London, New York and Paris', *Knowledge and Process Management*, 14(2): 117–30.

Sørensen, C. (2004) 'The Future Role of Trust in Work – The Key Success Factor for Mobile Productivity', Microsoft.

——. (2010) 'Cultivating Interaction Ubiquity at Work', *Information Society*, 26(4): 276–87.

——. (2011) 'Mobile IT', in B. Galliers and W. Currie (eds), *The Oxford Handbook of Management Information Systems: Critical Perspectives and New Directions*. Oxford University Press.

Sørensen, C. and Al-Taitoon, A. (2008) 'Organisational Usability of Mobile Computing: Volatility and Control in Mobile Foreign Exchange Trading', *International Journal of Human-Computer Studies*, 66(12): 916–29.

Sørensen, C., Al-Taitoon, A., Kietzmann, J., Pica, D., Wiredu, G., Elaluf-Calderwood, S., Boateng, K., Kakihara, M. and Gibson, D. (2008) 'Enterprise Mobility: Lessons from the Field', *Information Knowledge Systems Management Journal*, 7(1 and 2): 243–71.

Sørensen, C. and Gibson, D. (2004) 'Ubiquitous Visions and Opaque Realities: Professionals Talking About Mobile Technologies', *INFO: The Journal of Policy, Regulation and Strategy for Telecommunication, Information and Media*, 6(3): 188–96.

——. (2008) 'The Professional's Everyday Struggle to Ubiquitize Computers', in M. Elliott and K. Kraemer (eds), *Computerization Movements and Technology Diffusion: From Mainframes to Ubiquitous Computing*. Medford, NJ: Information Today, pp. 455–79.

Sørensen, C. and Pica, D. (2005) 'Tales from the Police: Mobile Technologies and Contexts of Work', *Information and Organization*, 15(3): 125–49.

Sørensen, C., Yoo, Y., Lyytinen, K. and DeGross, J.I. (eds) (2005) *Designing Ubiquitous Information Environments: Socio-technical Issues and Challenges*. New York: Springer.

Sproull, L. and Kiesler, S. (1993) *Connections: New Ways of Working in the Networked Organization*. Cambridge, MA: MIT Press.

Standage, T. (1998) *The Victorian Internet*. London: Weidenfeld & Nicolson.

Star, S.L. and Griesemer, J.R. (1989) 'Institutional Ecology, "Translations" and Boundary Objects: Amateurs and Professionals in Berkeley's Museum of Vertebrate Zoology, 1907–39', *Social Studies of Science*, 19: 387–420.

Star, S.L. and Ruhleder, K. (1996) 'Steps Toward an Ecology of Infrastructure: Design and Access for Large Information Spaces', *Information Systems Research*, 7(1): 111–34.

Stewart, I. (2007) *Why Beauty Is Truth: The History of Symmetry*. New York: Basic Books.

Straus, S.G., Bikson, T.K., Balkovich, E. and Pane, J.F. (2010) Mobile Technology and Action Teams: Assessing BlackBerry Use in Law Enforcement Units', *Computer Supported Cooperative Work (CSCW)*, 19(1): 45–71.

Strauss, A. (1985) 'Work and the Division of Labor', *Sociological Quarterly*, 26(1): 1–19.

Sturdy, A., Handley, K., Clark, T. and Fincham, R. (2009) *Management Consultancy: Boundaries and Knowledge in Action*. Oxford University Press.

Suchman, L. (1987) *Plans and Situated Actions. The Problem of Human-Machine Communication*. Cambridge University Press.

——. (1994) 'Do Categories Have Politics? The Language/Action Perspective Reconsidered', *Computer Supported Cooperative Work*, 2(3): 177–91.

——. (2006) *Human and Machine Reconfigurations: Plans and Situated Actions*. Cambridge University Press.

Sundaramurthy, C. and Lewis, M. (2003) 'Control and Collaboration: Paradoxes of Governance', *Academy of Management Review*, 28(3): 397–415.

Susskind, R.E. (2008) *The End of Lawyers?: Rethinking the Nature of Legal Services*. New York: Oxford University Press.

Tang, J.C., Isaacs, E. and Rua, M. (1994) 'Supporting Distributed Groups with a Mondtage of Lightweight Interactions', in T. Malone (ed.), *CSCW '94. Proceedings of the Conference on Computer-Supported Cooperative Work, Chapel Hill, North Carolina, October 24–26, 1994*. New York: ACM Press, pp. 441–52.

Taylor, M. (2007) 'Are You Addicted to Your "Crackberry"?', http://www.bmj.com/cgi/content/full/334/7587/241, date accessed 6 May 2011.

Thompson, J.D. (1967) *Organizations in Action. Social Science Base of Administrative Theory*. New York: McGraw-Hill.

Tilson, D., Lyytinen, K. and Sørensen, C. (2010) 'Digital Infrastructures: The Missing IS Research Agenda', *Information Systems Research*, 21(4): 748–59.

Tuan, Y.-F. (1977) *Space and Place: The Perspective of Experience*. London: Edward Arnold.

Turner, J.A. (1984) 'Computer Mediated Work: The Interplay Between Technology and Structured Jobs', *Communications of the ACM*, 27(12): 1210–17.

Turner, P. (2005) 'Affordance as Context', *Interacting with Computers*, 17(6): 787–800.

Urry, J. (2000) 'Mobile Sociology', *British Journal of Sociology*, 51(1): 185–203.

——. (2003) *Global Complexity*. Cambridge: Polity Press.

——. (2007) *Mobilities*. Cambridge: Polity Press.

Van de Ven, A.H., Delbecq, A.L. and Koenig Jr., R. (1976) 'Determinants of Coordination Modes within Organizations', *American Sociological Review*, 41(2): 322–38.

Van Fenema, P.C. and Kotlarsky, J. (2008) 'Information Technology for Personal, Impersonal and Automated e-Coordination Modes', in J. Kotlarsky, I. Oshri and P. Van Fenema (eds), *Knowledge Processes in Globally Distributed Contexts*. Basingstoke: Palgrave Macmillan, pp. 216–42.

Van Maanen, J. (1979) 'The Fact of Fiction in Organizational Ethnography', *Administrative Science Quarterly*, 24(4): 539–50.

——. (1988) *Tales of the Field: On Writing Ethnography*. University of Chicago Press.

Vincent, J. and Fortunati, L. (eds) (2009) *Electronic Emotion: The Mediation of Emotion via Information and Communication Technologies*. New York: Peter Lang Publishing.

Voss, C.A. and Hsuan, J. (2009) 'Service Architecture and Modularity', *Decision Sciences*, 40(3): 541–69.

Voutsina, K. (2008) 'IT Experts in Flexible Forms of Employment', unpublished doctoral dissertation, London School of Economics.

Voutsina, K., Kallinikos, J. and Sørensen, C. (2007) 'Codification and Transferability of IT Knowledge', in R. Winter and H. Österle (eds), *Proceedings of the 15th European Conference on Information Systems (ECIS)*. University of St Gallen.

Wajcman, J., Bittman, M. and Brown, J.E. (2008) 'Families without Borders: Mobile Phones, Connectedness and Work-Home Divisions', *Sociology*, 42(4): 635–52.

——. (2009) 'Intimate Connections: The Impact of the Mobile Phone', in G. Goggin and L. Hjort (eds), *Mobile Technologies: From Telecommunications to Media*. London: Routledge, pp. 9–22.

Waldman, D.A. and Yammarino, F.J. (1999) 'CEO Charismatic Leadership: Levels-of-Management and Levels-of-Analysis Effects', *Academy of Management Review*, 24(2): 266–85.

Warwick, K. (2002) *I, Cyborg*. London: Century.

Watson-Manheim, M.B. and Bélanger, F. (2007) 'Communication Media Repertoires: Dealing with the Multiplicity of Media Choices', *MIS Quarterly*, 31(2): 267–93.

Webb, W. (2010) *Being Mobile: Future Wireless Technologies and Applications*. Cambridge University Press.

Wegner, P. (1997) 'Why Interaction is More Powerful than Algorithms', *Communications of the ACM*, 40(5): 80–91.

Weick, K.E. (1998) 'Improvisation as a Mindset for Organizational Analysis', *Organization Science*, 9(5): 543–55.

Weilenmann, A. (2003) 'Doing Mobility', PhD dissertation, Gothenburg University.

Weilenmann, A. and Larsson, C. (2001) 'Local Use and Sharing of Mobile Phones', in B. Brown, N. Green and R. Harper (eds), *Wireless World*. Godalming: Springer-Verlag UK, pp. 99–115.

Weiser, M. (1991) 'The Computer for the Twenty-First Century', *Scientific American*, 265(3): 94–104.

Whitley, E.A. and Hosein, G. (2010) *Global Challenges for Identity Policies. Technology, Work and Globalization*. Basingstoke: Palgrave Macmillan.

Whittaker, S., Frohlich, D. and Daly-Jones, O. (1994) 'Informal Workplace Communication: What is It Like and How Might We Support It?', in B. Adelsom, S. Dumais and H. Olson (eds), *ACM 1994 Conference on Human Factors in Computing Systems*, Boston, MA: ACM Press, pp. 131–7.

Whittaker, S. and Sidner, C. (1996) 'Email Overload: Exploring Personal Information Management of Email', in R. Bilger, S. Guest and M. Tauber (eds), *ACM 1996 Conference on Human Factors in Computing Systems*. Vancouver: ACM Press, pp. 276–83.

Whittaker, S., Swanson, J., Kucan, J. and Sidner, C. (1997) 'TeleNotes: Managing Lightweight Interactions in the Desktop', *Transactions on Computer Human Interaction*, 4: 137–68.

Whittaker, S., Terveen, L. and Nardi, B.A. (2000) 'Let's Stop Pushing the Envelope and Start Addressing it: A Reference Task Agenda for HCI', *Human-Computer Interaction*, 15: 75–106.

Whittle, A. (2005) 'Preaching and Practising "Flexibility": Implications for Theories of Subjectivity at Work', *Human Relations*, 58(10): 1301–22.

Whittle, A. and Mueller, F. (2009) 'I Could Be Dead for Two Weeks and My Boss Would Never Know: Telework and the Politics of Representation', *New Technology, Work and Employment*, 24(2): 131–43.

Wiberg, M. (2001) 'Inbetween Mobile Meetings: Exploring Seamless Ongoing Interaction Support for Mobile CSCW', PhD dissertation, Department for Informatics, Umeå University.

Wiberg, M. and Ljungberg, F. (2001) 'Exploring the Vision of "Anytime, Anywhere" in the Context of Mobile Work', in Y. Malhotra (ed.), *Knowledge Management and Virtual Organizations*. Hershey, PA: Idea Group Publishing, pp. 157–69.

Wiberg, M. and Whittaker, S. (2005) 'Managing Availability: Supporting Lightweight Negotiations to Handle Interruptions', *ACM Transactions of Computer-Human Interaction*, 12(4): 1–32.

Williamson, O.E. (1981) 'The Economics of Organization: The Transaction Cost Approach', *American Journal of Sociology*, 87(3): 548–77.

Willmott, H. (1993) 'Strength is Ignorance; Slavery is Freedom: Managing Culture in Modern Organizations', *Journal of Management Studies*, 30(4): 515–52.

Winograd, T. (1994) 'Categories, Disciplines and Social Coordination', *Journal of Computer-Supported Cooperative Work*, 2(4): 191–7.

Winograd, T. and Flores, F. (1986) *Understanding Computers and Cognition: A New Foundation for Design*. Norwood, NJ: Ablex Publishing Corp.

Wiredu, G. (2005) 'Mobile Computing in Work-Integrated Learning: Problems of Remotely Distributed Activities and Technology Use', PhD dissertation, London School of Economics and Political Science.

——. (2007) 'User Appropriation of Mobile Technologies: Motives, Conditions and Design Properties', *Information and Organization*, 17: 110–29.

Wiredu, G. and Sørensen, C. (2006) 'The Dynamics of Control and Use of Mobile Technology in Distributed Activities', *European Journal of Information Systems*, 15(3): 307–19.

——. (Forthcoming) *Human Mobility in Distributed Organizing as a Coordination Process*. Manuscript in preparation.

Woods, D.D. (1988) 'Coping with Complexity: The Psychology of Human Behavior in Complex Systems', in L. Goodstein, H. Andersen and S. Olsen (eds), *Tasks, Errors and Mental Models. A Festschrift to Celebrate the 60th Birthday of Professor Jens Rasmussen*. London: Taylor & Francis, pp. 128–48.

Yates, J. (1989) *Control through Communication: The Rise of System in American Management*. Baltimore, MD: Johns Hopkins University Press.

Yoo, Y. (2010) 'Computing in Everyday Life: A Call for Research on Experiential Computing', *MIS Quarterly*, 34(2): 213–31.

Yoo, Y., Henfridsson, O. and Lyytinen, K. (2010) 'The New Organizing Logic of Digital Innovation: An Agenda for Information Systems Research', *Information Systems Research*, 21(4): 724–35.

York, J. and Pendharkar, P.C. (2004) 'Human–Computer Interaction Issues for Mobile Computing in a Variable Work Context', *International Journal of Human-Computer Studies*, 60: 771–97.

Zaloom, C. (2006) *Out of the Pits: Traders and Technology from Chicago to London*. Chicago University Press.

Zammuto, R.F., Griffith, T.L., Majchrzak, A., Dougherty, D.J. and Faraj, S. (2007) 'Information Technology and the Changing Fabric of Organization', *Organization Science*, 18(5): 749–62.

Zerubavel, E. (1981) *Hidden Rhythms: Schedules and Calendars in Social Life*. Berkeley, CA: University of California Press.

Zheng, Y., Venters, W. and Cornford, T. (Forthcoming) 'Collective Agility, Paradox and Organisational Improvisation: The Development of a Particle Physics GRID', *Information Systems Journal*.

Zolin, R., Hinds, P.J., Fruchter, R. and Levitt, R.E. (2004) 'Interpresonal Trust in Cross-functional, Geographically Distributed Work: A Longitudinal Study', *Information and Organization*, 14: 1–26.

Zuboff, S. (1988) *In the Age of the Smart Machine*. New York: Basic Books.

Zuboff, S. and Maxmin, J. (2002) *The Support Economy: Why Corporations are Failing Individuals and the Next Episode of Capitalism*. London: Penguin.

Author Index

Aaen, I.   171
Abowd, G.D.   24
Ackroyd, S.   103
Adomavicius, G.   141
Aducci, R.   98
Agar, J.   7, 18, 103
Ahuja M.K.   89
Al-Taitoon, A.   10–12, 60, 89, 93, 133, 171
Albrecht, K.   8, 27, 136
Alderman, J.   20
Aldrich, H.   96
Andersen, P.B.   17, 156
Andersson, P.   44
Andriessen, J.H.E .   44
Andriopoulo, C.   51–2
Angus, A.   8
Antero, M.   11
Aoki, M.   118
Arnold, M.   26, 32–3, 49, 52–5, 91
Arora, R.   11
Ashforth, B.E.   89
Athanasiou, T.   178
Avgerou, C.   6
Axtell, C.   43, 86

Baecker, R.M.   65
Baird, S.   134
Baldwin, C.Y.   141
Balkovich, E.   191
Bannon, L.   65–6, 122, 151
Baresi, L.   44
Barfield, W.   9, 25
Barley, S.R.   3, 33, 43, 58, 61, 74, 118, 153
Barnes, S.   1, 110
Baron, N.S.   7
Barrett, M.I.   42
Basole, R.C.   1, 8, 44, 142
Bassoli, A.   6, 8, 24, 46, 160
Batt, R.   73
Bauman, Z.   2, 47
Becker, F.   40
Beech, N.   51, 165
Bélanger, F.   64

Bell, D.   2, 142
Benford, S.   177
Beniger, J.R.   70, 164, 167
Bennett, J.   65
Berg, M.   181
Berghel, H.   20
Bikson, T.   191
Bilderbeek, P.   173
Birkinshaw, J.   2, 51–2, 167
Bish, R.L.   103
Bittman, M.   8
Bittner, E.   103
Blackler, F.   85
Bloodgood, J.   50–1
Boateng, K.   10–11, 13, 127
Bockstedt, J.   173
Bolter, J.D.   21
Boudreau, M.-C.   51
Braa, K.   183–4
Brewer, J.   174
Broadbent, M.   142
Brodie, R.J.   23
Brodt, T.L.   132
Brown, J.S.   33, 85
Brown, B.   92
Bruegger, U.   87
Bullen, C.   65
Bunting, M.   3, 41
Bunzel, D.   94
Burrell, G.   6, 41, 47
Burt, R.S.   189
Burton-Jones, A.   31, 141, 143
Büscher, M.   40

Cairncross, F.   46, 90
Cameron, K.S.   50
Caminer, D.   20
Carlstein, T.   6, 40
Carnoy, M.   3, 41
Carr, N.G.   81, 99
Carstensen, P.   29, 36, 115, 151
Casey, E.S.   39
Castells, M.   2–3, 7, 9, 19, 41, 44, 47–8

Subject Index

accountability 105–6, 122
action 6, 26, 36, 63
  emerging 60, 70–1, 75, 79, 85, 93,
    113, 127, 129, 152, 166, 168
ad hoc co-ordination 75, 109
ad hoc interaction 75, 162
addiction 98
affordances 15, 17, 31–7, 53, 57, 63, 71,
    86, 88, 116–17, 124, 148–9, 151–2, 168
  categories of 53, 148
  perceived 32–3
  technological 28, 52–3, 55, 81, 91
alerts 88–9, 93–4
algorithmic codifications 153
algorithmic encounters 142
algorithmic processing 22
algorithms 21–3, 153
alignment 129, 153–4
ambidextrous 2, 50, 160, 165, 169
application domain 68
architecture 6, 39
articulation work 68–9, 73–5, 163, 165
artifacts 32, 35–6, 45–6, 152
asymmetry 28–9, 37, 55–6, 115–16, 124,
    138–9, 143, 148, 163
asynchronous
  connectivity 4
  emails 80, 92
attention 16, 24, 26, 64, 67, 78, 80, 84,
    95–7, 99, 106–8, 154
awareness 25, 34, 55, 96, 109–10, 115,
    122, 124, 135, 143

barriers 58, 86, 98–100, 110, 115, 134
  collaborative 16, 108, 116, 125
  contextual 69
  geographical 61
  interactional 109
batch processing 115
batch system 114
Black Cabs 12, 82–4, 86, 91, 148
BlackBerry 80, 82
body 21, 25–8, 34, 63, 87, 105

body-machine relationship 26
boundaries 1, 4–6, 43, 47–8, 58–9, 85–6,
    89, 100–1, 109, 117, 134, 138, 160–1,
    164–5, 168
  blurring 137, 160
  collaborative 115, 162–3, 168
  designed 138
  disappearing 47
  fixing 100
  fluid 48
  individual 161–2, 168

cab drivers 43, 83–6, 93, 95, 107, 136,
    144, 159
carphones 18
cells 18, 25, 40, 171
centralised system 110, 113
closeness 54, 142, 147
co-construction 43, 46, 152
co-evolution 7, 49, 103, 109
co-ordination 3, 35, 43, 45–6, 54, 65, 69,
    78, 85, 87–9, 97, 101, 104, 106,
    108–11, 113, 116–17, 119, 122, 128,
    131, 133, 135, 138, 151–2, 162, 165–7
  distributed 135, 151, 162
  micro 123
co-ordination mechanisms 29, 35, 70,
    75, 113, 116, 124, 148, 151–2, 166
co-ordinator 130, 146, 148–50
cockpit 83–4, 86
collaboration 1, 34–7, 42, 45–6, 58–9,
    64–70, 72–3, 74, 100–1, 105, 107,
    109–10, 115–25, 137, 151, 160, 162,
    165
collaborative arrangements 58, 74,
    106–7, 115, 118–19, 120, 122, 163
collaborative practices 16, 117
collaborative work 66, 68–9, 73, 123
collaborators 5, 25, 29, 35, 42, 46, 57,
    65, 69, 156, 162
commodification 171
communication 80–1, 90, 106–7, 109,
    127, 131–2, 137–8, 172